Susu farmer broadcast-sowing rice, northern Sierra Leone. *Photo by Endre Nyerges.*

The Ecology of Practice

Food and Nutrition in History and Anthropology
A series edited by Solomon H. Katz, University of Pennsylvania

This book is part of a series. The publisher will accept continuation orders which may be cancelled
at any time and which provide for automatic billing and shipping of each title in the series upon
publication. Please write for details.

The Ecology of Practice

Studies of Food Crop Production in Sub-Saharan West Africa

Edited by

A. Endre Nyerges

Centre College
Danville, Kentucky

Routledge
Taylor & Francis Group

LONDON AND NEW YORK

First published 1997 by OPA (Overseas Publishers Association)

Published 2016 by Routledge
4 Park Square, Milton Park, Abingdon, Oxon OX14 4RN
605 Third Avenue, New York, NY 10017

Routledge is an imprint of the Taylor & Francis Group, an informa business

British Library Cataloguing in Publication Data

The ecology of practice: studies of food crop production
   in Sub-Saharan West Africa. - (Food and nutrition in
   history and anthropology; v. 12)
   1. Food crops - Africa, Sub-Saharan 2. Agricultural
   productivity - Africa, Sub-Saharan 3. Land use, Rural -
   Africa, Sub-Saharan - Social aspects
   I. Nyerges, A. Endre
   338.1′967

ISBN 13: 978-90-5699-574-4 (pbk)
ISBN 13: 978-1-315-07839-7 (ebk)

For my father
Anton N. Nyerges, PhD
1917–1989

Contents

Series Editor's Preface

The study of the origin, development and diversity of the human diet is emerging as a coherent field that offers a much-needed integrative framework for our contemporary knowledge of the ecology of food and nutrition. This authoritative series of monographs and symposia volumes on the history and anthropology of food and nutrition is designed to address this need by providing integrative approaches to the study of various problems within the human food chain. Since the series is both methodologically and conceptually integrative, the focus of the individual volumes spans such topics as nutrition and health, culinary practices, prehistoric analyses of diet, and food scarcity and subsistence practices among various societies of the world. As a series, it offers many unique opportunities for a wide range of scientists, scholars and other professionals representing anthropology, archaeology, food history, economics, agriculture, folklore, nutrition, medicine, pharmacology, public health and public policy to exchange important new knowledge, discoveries and methods involved in the study of all aspects of human foodways.

Solomon H. Katz

Acknowledgments

This book had its origin in a dinner-table conversation with Art Hansen in Gainesville, Florida, in the late 1980s. Hansen suggested that I edit a book on rice in Africa but also urged me to work on whatever project was most interesting to me. In 1991, I organized a session at the annual meeting of the American Anthropological Association on the topic "The Ecology and Economics of Food Crop Production in Sub-Saharan West Africa," under the co-sponsorship of the Culture and Agriculture Group and the AAA Task Force on African Hunger, Famine and Food Security. Some of the panelists agreed to pursue a book project, and I began the effort to produce a volume that would coalesce some of the burgeoning interest in and knowledge of western African ecology and agriculture.

For encouragement, I owe a sincere debt of gratitude to Sol Katz and Rebecca Huss-Ashmore and, most especially, Olga Linares, whose patience and positive attitude kept me going forward on a project that sometimes seemed endless.

Barbara Cellarius, Caroline Dunn, and Nina James-Fowler contributed valuable editorial assistance. Scott Stauber is responsible for the illustrations and Margaret Altman did the index and helped with the proofreading. Students in my seminar on ecological anthropology, spring 1996, read and thoughtfully critiqued the manuscript. Criticisms by Gregg Goldstein were particularly apt. Rhonda Gillett-Netting kindly gave me permission to quote extensively from Robert McC. Netting's unpublished "Comments."

The College of Arts and Sciences, University of Kentucky, in Lexington, generously supported this project with research assistantships and a Summer Faculty Research Fellowship.

I am grateful to Scott Bentley, who nurtured the project in its early stages, and to Carol Hollander, Janell Robisch and S. Ananthalakshmi, who competently saw it through to its conclusion. I thank my mother Helen Nyerges for years of support and Lee Blonder and Eva Nyerges for years of patience.

Finally, Lucas Nyerges kept me company through many weekend hours, playing quietly in my office while I worked on writing and revising the manuscript.

CHAPTER 1

Introduction–The Ecology of Practice

A. Endre Nyerges
Anthropology/Sociology Program
Centre College, Danville, Kentucky

This introduction has two aims: to preview the chapters that follow and to propose and briefly outline a practice perspective in ecological anthropology. This idea can be expressed as follows. Humans exploit resources for social purposes and in the context of competitive, often hierarchical social arrangements. These purposes and the cultural goals and values behind them create the conditions for the individual use and modification of resources, as people attempt to meet their basic needs. Any study in human ecology must take into account these conditions, and, therefore, it is imperative to ground analyses in ecology in individual activity appropriately culturally contextualized.

This perspective, although by no means universally accepted (nor totally new), exemplifies a basic trend in anthropology away from simple reductionism and "vulgar" materialism. It resolves, or is intended to resolve, a number of perceived weakness in what should be now, because of its intrinsic interest and merit, one of the most vigorous and important fields in cultural anthropology as we "pick up the pieces and somehow resume the tasks of regular anthropology" (Barth 1994:349) following the decade of critique. The chapters to follow are individually authored ecological case studies concerning food production in sub-Saharan Africa, in which the themes and ideas of the introduction are represented. Overall, the intent of the book is to present a sociocentric approach in ecology – one that is focused on conflict and power relations among groups and individuals of different statuses in society (Spooner 1982, 1984) – providing examples from field studies, and, even more importantly, perhaps stimulating new field work. My hope is that the book, and this introduction with it, will encourage others to critique and further clarify the practice paradigm, which holds that it is incorrect to treat resources and environments as if they were somehow extrinsic to and generative of production systems. Rather, individual agency (practice, politics) links the exploitation of resources to technologies that are created and used for the realization of culturally important projects. In the remainder of this introduction, I discuss the development of ideas in ecological anthropology, focusing on recent concerns and formulations in the field, suggesting ways in which a practice perspective is a natural outgrowth of these concerns, and conceptualizing some of the features of a practice approach. I then turn to the task of previewing the book's substantive chapters in the context of the theoretical discussion.

METHOD AND THEORY IN ECOLOGICAL ANTHROPOLOGY

From Ecosystems and Populations to Resources and Actors

In recent decades, ecological anthropologists have worked in the absence of a general theoretical and methodological agreement. This has not always been the case, and perhaps we can look forward to a day when that again will change. Not too long ago in the sixties and early seventies, however, ecological anthropologists were concerned with a limited number of related research themes, typically focused on ecosystems. In the ecosystems approach, human groups were considered as equivalent to natural populations (Vayda and Rappaport 1968), and cultural behavior was defined in terms of its functioning to regulate energy flow (following Rappaport 1968 [1984]). In research based on this approach, ecological anthropologists examined the ways in which culture served to maintain human populations in resource (i.e., calorie) balance with the natural environment.

At its best, this work focused on cultural adaptations as the means to establish, restore, or reestablish an ecological equilibrium once a former balance was lost (Orlove 1980:240–241). Accordingly, certain features of culture and society, such as ritual, were conceived as having the "latent" or invisible function of ecosystem maintenance through acting as cybernetic regulators. Thus, in *Pigs for the Ancestors*, Rappaport (1968 [1984]) described a ritually regulated ecological cycle among the Tsembaga Maring of New Guinea. In this society, an increase in the population of feral pigs over a period of 15 to 20 years caused a decline in garden yam yields, eventually leading to the initiation of a period of pig slaughter and consumption in the feasting associated with warfare. The decline in the number of pigs during the feasting part of the cycle then restored the original, lost balance between the biomass of yams and the populations of pigs and humans that fed on them. Yet, at the ritual close of the warfare period pig consumption would cease, the pig population would recover, and the cycle would begin anew.

In addition to this perspective, there has also existed in anthropology a strongly deterministic or nomothetic position, i.e., the "science of culture" of Marvin Harris's *Cultural Materialism* (1979; see also Murphy and Margolis 1995). This view explains human cultural evolution largely through an infrastructural core of "demo-techno-econo-environmental" factors, including the mode of production and mode of reproduction, which forms

> the principal interface between culture and nature, the boundary across which the ecological, chemical, and physical restraints to which human action is subject interact with the principal sociocultural practices aimed at overcoming or modifying those restraints. (Harris 1979:57)

The restraints are ultimately based on laws of nature that can never be altered. Hence, in Harris's formulation the infrastructure is accorded a "strategic priority" for theory building, due to its role in probabilistically determining the rest of culture and society, i.e., the social and political "structure" and ideological and symbolic "superstructure." An example of this approach is Harris's (1977:Ch. 1) now-famous study of the sacred cow in the history of agricultural intensification in the Indus region. In this process, a period of long-term environmental decline brought about over thousands of years of overgrazing by cattle herded for meat led, ultimately, to the evolution of a system of ritual protection of the cattle exploited primarily as draft animals in the production of food crops. This change, accomplished by the

sacralization of cattle, resulted in the establishment of a new ecological balance between plant biomass, on the one hand, and populations
of cattle and humans, on the other.

In general, then, the approaches characterized by the work of
Rappaport and Harris borrowed key concepts (e.g., ecosystems, adaptation) from biological ecology and focused on issues of homeostasis
and change. Hallmarks were the materialist explanation of cultural
"riddles" (e.g., Harner 1977) and the production of illustrative dia-
grams that purported to measure the amount and flow of calories
through a culturally regulated and energetically balanced ecosystem
(e.g., Rappaport 1971). A strength of both approaches was their holism.
A basic contrast was the locus of causation: for ecosystems ecology the
entire sociocultural system, in particular ritual as sanctified practice,
functioned as "a regulator or information control filter promoting homeostasis or social change" (Patterson 1987:27); for cultural materialism, "[n]ature in the form of demographic and ecological pressures...
was ultimately responsible for stability and change" (1987:27).

In addition to these approaches, other less deterministic, but
nonetheless still strongly materialist, positions have included the recognition of a progressive trend in societies from a state of autarky
and dependence on local natural resources at low population densities
to more complex states of intensification of resource use, greater control over resources, and dependence on broader economic linkages at
higher densities. Given this shift, called the "ecological transition" by
Bennett (1976, 1993:3–11), concepts of ecological adaptation, e.g., optimal foraging theory, can be readily applied in studies of Eskimo and
other hunter-gatherer foraging behavior (i.e., Winterhalder and Smith
1981; Smith 1991), and in those horticultural societies that, like the
Machiguenga of Peru, can be described as adapted in part because
they lack any superordinate political structure beyond the level of the
household (Johnson 1983). In the case of chiefdom and state societies,
however, a recent examination of ethnographic, ethnohistorical, and
archaeological knowledge suggests a lack of correlation between technological, demographic, and social factors, on the one hand, and political organization, on the other (Netting 1990). In examining such
societies, therefore, evolutionary concepts including adaptation can
appropriately be applied to the levels of technology and the organization and functioning of the family or household, but not to higher-
level sociocultural or political orders.

This approach is ultimately based on the cultural ecology of Julian
Steward (1955), in combination with the population growth correlates

of Ester Boserup (1965). It is restricted in scope, on the one hand, to environmental features that serve as resources and, on the other, to changing aspects of culture and society – the "cultural core" – that are most immediately implicated in technology, demography, and the organization and mobilization of labor. Researchers working in this tradition seek to examine

> a limited set of social and economic factors that are regularly associated with a definable type of productive activity, despite considerable variation in a number of other "cultural" features that may themselves cohere internally in meaningful, patterned ways. (Netting 1993:2)

The approach is exemplified by works such as Netting's recent *Smallholders, Householders* (1993), a study of sustainable farming by smallholders in varying contexts of population densities and changing domestic organizations. Generally, work based on the "microperspective" of cultural ecology highlights the technological successes of local farming systems, e.g., practices of terracing, ridging, mulching, polyculture, and household strategies of labor organization and mobilization, as they are applied in fulfilling basic human needs (Richards 1983; Huss-Ashmore 1989:27–28).

For many current practitioners, then, ecological approaches in anthropology have shifted in recent years from a dominant focus on total systems in balance to less deterministic perspectives that emphasize links between demography, domestic organization, agricultural intensity, and technology. Increasingly, ecosystems are characterized as not balanced (Botkin 1990; Vayda 1991), and ecological anthropologists now regularly study productive activities "without necessarily attempting to show how they maintain populations in equilibrium" (Orlove 1980:249), particularly with ecological cycles of calories, other nutrients, or other physical or abiotic features of the environment. The research is particularistic and characteristically focused on technological knowledge and "coping," as for example on the exploitation of multiple varieties of rice in an innovative indigenous agricultural system (Richards 1985).

Similarly, the concept of aggregate human populations in adaptation has given way to an emphasis on smaller-order structures. These include communities, households (e.g., Guyer 1981; Netting, Wilk, and Arnould 1984), and, finally, individual humans as social actors (Vayda and McCay 1975; McCay 1978; Orlove 1980; Vayda 1983; Lees and Bates 1990), with each emphasizing, as Spooner (1987:64) states, "the convenience of the family group in the context of its social matrix."

Thus, Bennett's "adaptive strategies" approach concerns the "rational or purposive manipulation of the social and natural environments" by human actors in the context of the "ecological transition" – i.e., "the progressive incorporation of Nature into human frames of purpose and action" (Bennett 1976:3; reviewed in Orlove 1980: 246–247, 251). Bennett's perspective includes concepts of culturally defined needs, the creation of strategic actions by "voluntary or *sui generis* forces" (1976:165), and the increasing tendency for the emergence of hierarchies of status and power (1976:6). This approach, which focuses on the adaptive strategies and coping mechanisms of individuals, can be summed up neatly in the title of Bennett's recent book: *Human Ecology as Human Behavior* (1993).

This paradigmatic shift, from ecosystems and populations to resources and actors, not only brings ecological anthropology closer in line with theory in evolutionary ecology, which emphasizes individuals, but also reflects a general move in anthropology "from social structure to social process, from treating populations as uniform to examining diversity and variability within them, and from normative and jural aspects to behavioral aspects of social relations" (Orlove 1980:246). As is generally recognized, this view is grounded in Firth's (1951:28–40) distinction between *social structure*, as an ideal pattern of relations or enduring expectations about people's behavior in society, and *social organization*, as the acts and choices of individuals that change or maintain the structures. Avenues of research that stem from this focus – on organization as opposed to structure, acts as opposed to rules – include the actor-centered, action, and transactional approaches originally developed in social and political anthropology (Barth 1959, 1966, 1967; Swartz, Turner, and Tuden 1966; Bailey 1969; summarized in Lewellen 1992), and further developed in studies of rational decision-making (Barlett 1980).

In ecological anthropology, perhaps the most powerful and persuasive statement of the actor-centered approach is the method of "progressive contextualization" espoused by Andrew Vayda (1983). In advocating this approach, Vayda extols the practical virtues of research that explicitly avoids any a priori definitions of social units and structures or their characteristics. Instead, he seeks to study and explain people-environment interactions by first "focusing on specific activities, such as timber cutting, performed by specific people in specific places at specific times" and only then "trac[ing] the causes and effects of these activities outward" (Vayda 1983:266). Progressive contextualization means "placing [human actions] within progressively wider or denser contexts" (Vayda 1983:265), but never defining the boundaries or other characteristics of human groups as involving "unwarranted assumptions." The empirical advantages of this actor-centered approach are evident. An investigator can begin with a particular resource user and simply proceed as far as time and money will allow. Moreover, specific social units such as tribes, communities, or ethnic groups, which may be permanent features of local society but nevertheless irrelevant to the ecological issue at hand, can safely be ignored in favor of the ad hoc coalitions of resource-users that are perhaps more characteristic of the contemporary context (Vayda 1986:286). Finally, in carrying out a program of progressive contextualization, the researcher is unimpeded in freely investigating the full range of actual behavioral variation in resource use, which clearly needs to be understood as a fundamental aspect of human life and one that is often obscured by predictive models of standard or normative behavior (Vayda 1994).

Yet, just as it would be inadequate to study human action apart from the environment in which it occurs ("the ecosystem and decisions made by individual actors affect each other reciprocally" – Orlove 1980:248), it would be equally inappropriate to portray human action, at any level of analysis, apart from its encompassing social and cultural contexts. Thus, a potential problem for an actor-centered ecology is a tendency to view the individual as apart from her or his social context, to relegate culture and society to the position of backdrop or stage setting for analytically prior individual action or decision-making. A general conclusion that might be reached, therefore, is that a better sociology needs to be attached to an actor-based model in ecology, in order to develop a more adequately sociocentric approach (Spooner 1984).

The Practice Orientation

A potential, albeit partial and presumably transitional, solution to the problems of integrating social and cultural factors into considerations of ecology is offered by the development of a practice orientation in anthropology (Ortner 1984; Karp 1986; Vayda 1986:299–300). According to practice theory (Bourdieu 1977 [1972]), or the theory of "structuration" (Giddens 1979, 1984), individuals are active operators in creating or shaping the social and cultural contexts that simultaneously frame or constrain their actions and decisions. Practice theory, which focuses on the continuous flow of activity in space and time, emphasizes the essential recursiveness of everyday social life, in which structure is both medium and outcome of the reproduction of practices. As Ortner (1984) states,

> practice theory seeks to explain the relationship(s) that obtain between [individual] human action, on the one hand, and some global entity which we may call "the system" on the other. Questions concerning these relationships may go in either direction – the impact of the system on practice, and the impact of practice on the system. (1984:148)

Practice theory thus characterizes individual action especially in terms of mundane, routine, or incompletely articulated activities, and further emphasizes that social relationships are inherently, and almost inevitably, asymmetrical. The practice orientation, then, includes a view of humans as active participants in and creators of culture and social life and, concomittantly, engaged in a social conflict over resources (Giddens 1979:150–155).

In *Central Problems in Social Theory*, Giddens (1979) discusses this theoretical shift in the social sciences from a focus on structure and aggregates to individuals and actors actively involved in modifying, or even creating, "structure." He links this discussion of individual action to the social competition for resources and supports a materialist philosophy of history that includes, from Marx:

> a *conception of human Praxis*, emphasizing that *humans are neither to be treated as passive objects, nor as wholly free subjects.* The materialist conception of history, in this context, is opposed both to idealism and to "passive" or "mechanical" materialism. The most famous and brilliantly succinct exposition of this is given in the theses on Feuerbach, where Marx argues that *the main shortcoming of previous forms of materialism (and many subsequent types also, one might add!) is that the relations between human actors, and between human actors and the material world, are treated as ones of passive contemplation, not as active, practical relations.* The study of human life, Marx emphasizes, is the study of definite social practices, geared into human needs. The interaction of human beings with nature is one of active appropriation: 'The whole of history is a preparation for "man" to become an object of sense perception, and for the development of human needs (the needs of man as such).' (Giddens 1979:151–152; emphasis added)

Practice theory, then, unites a concern for individual human actors in their "active appropriation" of resources with an interest in the mediation of this relationship through "definite social practices, geared into human needs." Giddens describes his work as a "*nonfunctionalist manifesto*" (1979:7), in which he rejects functionalism's tendency to "get behind" the backs of its subjects and examine the unintended consequences of action (that is, how basic needs are fulfilled without people being explicitly aware of what they are actually doing, and why). Instead, he prefers to consider persons as actively involved in the systems of subordination and domination – and resource competition, control, and exploitation – in which they live. Moreover, they are explicitly knowledgeable of these systems and therefore capable of being agents in, rather than merely products of, them (Karp 1986). Additional key aspects include ideology, which is "one type of resource involved in domination" (Giddens 1979:6), and the role of ideological conflict by which sectional interests are (mis)represented as the interests of the whole group.

Practice theory makes intuitive sense, yet translating it into a research program for ecological anthropology (as others have advocated) may be difficult, and this approach has produced some "initial forays" only (Palm 1990:76–77). According to Palm's review of social science perspectives on human responses to natural hazards:

> An increasingly popular trend in the social sciences has been a position that combined the strengths of the structural determinist position with the perspectives obtained from a focus on individual behavior or volunteerism. Proponents of this philosophical framework reject a deterministic influence of either the structure or the individual (agency), but instead seek to interpret their relative and mutual roles within the context in which they have developed.... The impacts of influential individuals (agents), the influence of structure as translated by all individuals in their everyday lives into action and changes in the structure, and the influence of environmental structure, all are interwoven to describe particular responses to environment. Both the preexisting societal structure and the influence of individuals on and within this structure are investigated. (1990:76)

The interactions of preexisting social relationships and the behavior of the individual are at the heart of practice theory. I would, however, suggest that perhaps the ultimate significance of this perspective lies in its focus on the mundane and repetitive aspects of daily life, which also constitute the foci for resource use, and its emphasis on hierarchy, both in terms of how hierarchy is institutionalized and maintained in practice and how hierarchy in social life affects individual practice and resource use. Thus, hierarchies are institutions and therefore analyzable as "practices which are deeply sedimented in time-space" (Giddens 1979:80). For ecology, the key implication of this view is that, depending on their position within an established social order, members of society may respond to environmental factors variably and, therefore, manage resources differently.[1]

The Social Life of Resources

In formulating a practice approach in ecological anthropology, a question that necessarily arises is how individual action and hierarchical structuring actually affect production and resource use in specific research contexts. To begin to answer this question, the close connection of resources to individual action needs to be recognized, as suggested by the term "socionatural" systems coined by Bennett (1993:11–21) in an effort to escape the prevalent dualisms of a Nature–Culture dichotomy. Similarly, in previous literature concepts of "agroecosystems" (e.g., Janzen 1973) and "human use systems" (di Castri 1976) have been used to imply the social or geographical units through which environments are exploited. Although useful, these terms do not fully convey the explicitly *sociocultural* character of interactions between individual humans and the environment, i.e., the crucial concept of the *incorporation* of natural resources into the social lives of the individuals who exploit them.

In the volume *The Social Life of Things*, Appadurai (1986) and Kopytoff (1986) argue that just as each individual has a biography of her or his social life, so, too, do the material objects that individuals use and value. A prime example from the history of anthropology is the kula exchange system of the Trobrianders, as described by Malinowski in *Argonauts of the Western Pacific* (1922). In this study, the subject of analysis is, perhaps surprisingly, not the social life of the Trobrianders per se, but rather this life as understood through the Trobriander's material goods. These include the kula – the famed shell armbands and necklaces that circulate and function in Trobriand society and increase in individual value through the process of exchange. In other words, *things* such as exchange goods may be thought of socially, as having cultural biographies of the histories of their use, and as differentiated individually. As Kopytoff states, this view is based on "the profound Durkheimian notion that a society orders the world of things on the pattern of the structure that prevails in the social world of its people" (1986:90).[2] In Appadurai's formulation, the social life of things most importantly involves their place in competitive "tournaments of value," in which the exchange of luxury commodities creates and maintains elites (1986:20–23). This linking of individuated things to hierarchies through the cultural construction of value is obvious and has recently been elaborated on by Ferguson (1992) in a discussion of the "cultural topography" of wealth in Lesotho.

The concept of the cultural biography of a thing and of a socially determined history of its use can be applied, not to luxury goods only, but also to "'primary' or 'bulk' commodities" (Appadurai 1986:6). As Appadurai notes (1986:13) "[t]hough the biographical aspect of some things (such as heirlooms, postage stamps, and antiques) may be more noticeable than that of some others (such as steel bars, salt, or sugar), this component is never completely irrelevant." Indeed, there is a biographical or "social life" aspect to natural resources, as well, and although perhaps not explicitly stated as such the idea of resources as actively appropriated into social life has long been familiar to students of swidden agriculture. Thus, in a seminal paper, Conklin (1961: Figure 1) diagrammed a system of shifting cultivation as involving the interaction of sociocultural with biotic, edaphic, and climatic factors in the construction of a swidden site over time. This previous work, however, focused more on overall patterns in the ecology of resources and their exploitation, and less on variation. A true practice perspective, if ever actually implemented, would require explicitly distinguishing among resources in terms of the biographies of their creation, use, and incorporation by individual, hierarchically differentiated resource users as engaged in social life, i.e., in continuous, ongoing conflicts to maintain or change their social standing through resource exploitation.

The methodological implications of the ecology of practice are to distinguish actors according to social status, to examine access to and control over the means of production, and to show how conflict over control has consequences for the exploitation and management of specific resources as they are incorporated into individual social lives. In this way, it should be possible to integrate findings based on a "sociocentric" perspective and informed by considerations of everyday practice and hierarchy, with findings based on a "technocentric" approach and informed by concepts of adaptation or coping. The ecology of practice can, therefore, best be understood as an attempt to approach the study of "a species that lives, and can only live, in terms of meanings it itself must construct in a world devoid of intrinsic meaning but subject to natural law" (Rappaport 1994:154). To summarize, this approach

• rejects the ecosystems ecology position of viewing humans and human sociocultural behavior as mere passive regulators in systems of calorie flow (although accepting as fundamental the ecosystemic view that all life forms and the edaphic environment are intensely interconnected);

- takes as its crucial methodological starting point the individual actor exploiting resources;
- analytically privileges the sociocultural contexts in which individuals are acting, viewing these contexts as sometimes readily manipulable and sometimes highly constraining particularly when institutionalized into formalized structures, or hierarchies, of age, gender, ethnic, and class relations; and
- highlights the salient implications of viewing "natural" resources as things that are fully part of social life.

Illustrations of this approach can be found in the case studies to follow. These studies also exemplify aspects of the cultural ecology approach. In addition, as studies of food production in small-scale societies in the overall context of underdevelopment in western Africa, these cases share themes in common with studies of underdeveloped societies in general. These commonalities, a function of the incorporation of African states into a world economic system focused on Western nations, can best be understood through the conceptualizations of the field of political ecology. In the following section, I turn briefly to a consideration of the strengths of the political ecology approach and suggest how concepts of the social life of resources and the ecology of practice articulate with it.

Political Ecology

Political ecology, described as representing "a confluence between ecologically rooted social science and the principles of political economy" (Peet and Watts 1993:239), has recently emerged as a predominant model in ecological anthropology. Beginning with some early formulations in the 1970s (Wolf 1972), political ecology has come to reflect the understanding that local ecology has global links, not only through ecosystemic functioning (e.g., the greenhouse effect, global warming) but also through economic forms that now integrate all places and peoples in a world-wide capitalist system.

These ideas have been propounded in books such as Blaikie and Brookfield (1987) and Little and Horowitz, with Nyerges (1987), which explicitly link global processes, smallholder poverty, and local-level problems of food insufficiency and environmental decline. More recent examples are Stonich (1993), Greenberg and Park (1994), Painter and Durham (1995), and the papers presented at a session on "Political Ecology: Struggles to Redefine Human Environment Relations" held at the 94th Annual Meeting of the American Anthropological Association in November 1995. Although their specific content is variable, a theme common to all these works is the emphasis on integrating investigations of local-level human-environmental relations with external political and economic factors, often via the decision-making efforts of the local land manager (e.g., Blaikie and Brookfield 1987). In this context, Greenberg and Park's (1994) integrative essay, about "the relations between human society, viewed in its bio-cultural-political complexity, and a significantly humanized nature" (ibid.:1), is particularly helpful in portraying a world in which people interact over resources in ways that are, within limits, knowable and, simultaneously, conflict-ridden.

Political ecology emerges from the familiar critique of previous approaches as ahistorical and homeostatic (Watts 1983, 1984; Ortner 1984:134). For example, one author explicitly criticizes the "old ecology"

> for neglecting the political dimensions of human/environment interactions, and thus for treating human communities as if they were fairly homogeneous, autonomous isolates, adapting – or sometimes failing to adapt – to a given exogenous environment. From the cultural ecology of Julian Steward (1955) through the ecosystems ecology of Roy Rappaport (1968) and on to the cultural materialism of Marvin Harris (1979), the primary focus was upon mechanisms of population adjustment to the natural environment. The importance of political dynamics – both those internal to populations, as may be fundamental to differential access to resources within the aggregate, and those between local populations and the wider world – did emerge as a theme of a few specific works ... but these few case studies fell short of providing analytical tools worthy of a general new approach. A truly "political ecology" has only begun to emerge in the 1980s. (Durham 1995:249)

Political ecology has important implications for anthropological studies of resource use, and its value for policy formulation and for focusing attention on issues such as famine, food insecurity, and environmental change is evident, but it has also been critiqued. In particular,

> *political* ecology has very little politics – there is no serious attempt at treating the means by which control and access of resources or property rights are defined, negotiated, or contested within the political arenas of the household, the workplace, and the state – and they adopt a rather

old-fashioned view of ecology rooted in stability, resilience and systems theory. (Peet and Watts 1993:239)

This critique further suggests that "political ecology seems grounded less in a coherent theory than in similar areas of inquiry" and that "its theoretical coherence . . . remains in question" (Peet and Watts 1993:239). Similarly, Durham (1995:262–263) concludes his discussion of political ecology by indicating that "we need far more attention to cultural variables. . . . *And we need far better ways to bring inequality in all its guises – race, class, gender, and ethnicity – into the picture*" (emphasis added).

In this context, the present argument regarding the ecology of practice and the social life of resources should be seen as an attempt, not to dismiss or replace political ecology (thereby contributing further to the widespread and lamentable practice of "intellectual deforestation" – Wolf 1990: 588, cited in Greenberg and Park 1994:1), but rather to expand and reorient it in particularly important ways. Thus, Collins (1992) usefully indicates that many of the perceived weaknesses of the political ecology approach could be resolved through a renewed attention, following Julian Steward, to the labor process appropriately considered in its social and culture historical context. In my own work, I emphasize individual, local-level interactions in relation to natural resources and ecology, in which labor organization mediates between the technology of production and the social factors that sometimes constrain people's ability to successfully carry out their environmental adaptations.

Finally, a salient point of difference remains between the ecology of practice and political ecology approaches, and this merits explicit statement: political ecology is policy-driven and grew directly out of the perceived need, and consequent effort, to make the social sciences more relevant (Blaikie and Brookfield 1987:Ch. 1). I support this aim, but in defining concepts of the ecology of practice and the social life of resources, I primarily hope to prod ecological anthropology back in a direction that will once again place research on the mundane activities of natural resource management, viewed in appropriate local and regional ethnographic and culture historical contexts, at the forefront of theoretical developments in the discipline of anthropology.

CASE STUDIES

At this point, some examples might help in explicating the practice approach. To start hypothetically, the following description of the social life of the Kabre of Togo, West Africa, in the context of a study of secrecy (Piot 1993), provides an ideal beginning on which to base a case study in the ecology of practice of this society. As Piot informs us:

> [w]hile I would not argue for a transactional view of Kabre culture – as constructed out of the needs and strategic interactions of individuals – much of the stuff of daily life is nevertheless tied into the struggles and negotiations of individuals. In building exchange relationships, in dealing with witchcraft accusations, in complying with the demands of seniors, and, more generally, in realizing the long-term cultural projects of the society (building a following, establishing a "name," etc.), individuals are continually challenging and confronting one another.... Moreover, these struggles are informed by, and infused with, a value that is central to Kabre culture – that of hierarchy. Social relationships, Kabre say, should be hierarchical. Without "respect" (*nyamto*) or hierarchy, relationships are not orderly; people behave "any which way" (*yim*). Thus, for example, relationships between communities are hierarchical, as are those between parents and children, elders and juniors, men and women, ritual moities, and so on. (Piot 1993:356)

With this as a beginning, it should be possible to ask questions about, and investigate how, individual, hierarchically differentiated Kabre exploit the resources of their environment in order to succeed in the struggles and negotiations of daily social life and move forward in their cultural projects. Before proceeding to the investigation of actual patterns of Kabre production and resource use, we would know that individual Kabre can be expected to be different from one another on the basis of age, gender, and residential groupings, and a past history of individual successes and failures in specific social competitions.

We need not, however, rely on hypothetical examples to draw out the research implications of the ecology of practice, as the case studies in this book provide illustrations of many of the aspects of this approach, as well as of the others (political ecology, cultural ecology) that have been discussed here. In the materials to follow, the specific themes which inter-link the case studies of food crop production in western Africa south of the Sahara include:

- The human occupation of a West African environment of increasing rainfall, and associated vegetational change and changing agricultural potential, from the north (Sahara) to the south (Sahel, Sudan, Guinea savanna, and coastal forest). This environment is characterized throughout by problems of disease, drought, and seasonality (e.g., Richards 1986; Nyerges 1989; Linares 1992; Park 1993).
- A regional culture history in the westernmost portion of this area of a Mande diaspora from savanna to forest and associated processes of

agricultural, technological, and environmental change (e.g., Brooks 1993; Nyerges 1996). This process is coupled with a variable history of Islamic influence from the north and interior, and Euro-Christian control and influence from the south and coast (Ottenberg 1984).

- As a result of this history, societies of this region emphasize the subsistence production of Sudanic grains such as sorghum, millet and, above all, rice (Richards 1986:4–5), a crop which recently has also taken on new importance in the alimentation of regional urban populations.
- As a further consequence of this regional history, societies of western Africa experience economic peripheralization in world systems, with consequent implications for food insecurity (Little and Horowitz, with Nyerges 1987). Thus, problems of hunger and famine may be linked, not only to drought or seasonality, but also to threats of international development (Nyerges 1987; Horowitz 1991) and regional warfare (Richards 1992; Magistro 1993).
- Finally, many rural societies of this region exhibit a form of organization, essentially pre-colonial in origin, that may be described as a lineage mode of production (Meillassoux 1981 [1975]) or an economy of wealth-in-people (Kopytoff and Miers 1977; Kopytoff 1987). In this system the goals of production are directed to the accumulation of dependents and followers, both as agricultural producers (Nyerges 1992) and as stores of technological and esoteric knowledge (Guyer 1993; Guyer and Belinga 1995).

These factors are varyingly manifested in the different study areas and come to the fore in different ways and degrees in the case study chapters, as analyzed in the following section. The analysis interweaves examination of the studies in the regional context and in terms of the ecology of practice. It includes, where available and appropriate, the original discussant's comments by the late Robert McC. Netting (Netting 1991), made on a subset of the papers as presented at the 1991 meeting of the American Anthropological Association, in a session organized by the book editor on "The Ecology and Economics of Food Crop Production in Sub-Saharan West Africa."

The Jola of Lower Casamance

In "Diminished Rains and Divided Tasks: Rice-Growing in Three Jola Communities of Casamance, Senegal," Olga Linares examines the social organization of agricultural production in the context of climatic uncertainty and ecological stress. The analysis is based on the varying social dynamics that shape agricultural practices in Jola villages and focuses on the ways in which drought and the highly seasonal rainfall regime of the savanna environment affect production. Specifically, Linares compares three villages that are similar in size and involvement in rice production but that differ in part due to the historical influences of Mande culture, Islam, and European intrusion on lifeways and the social division of productive labor. She shows that, although the three communities were similar in terms of labor invested in agricultural production per hectare during the 1981 drought year, one (Jipalom) had significantly lower returns to labor (by about half) than the other two (Sambujat, Fatiya).

According to the analysis, this shortfall was the result, not of local differences in the amount of rainfall or the technology of rice production per se, but rather of problems in the timely mobilization of agricultural labor caused by the prevailing household division of labor. This system, in Jipalom, is fully cooperative, with men and women participating equally in the work demanded by farming fields of both rice and groundnuts/millet. As a result, each farming task (soil preparing, planting, weeding, harvesting) must be completed in two sets of fields before the next may begin, and the rate of progress through the agricultural cycle is consequently slowed. Moreover, in Jipalom in 1981, the socially important need to take part in a circumcision ritual in a neighboring village where many residents had uterine kin superseded the demands of the farming system, delaying the start of the agricultural year, and further exacerbating the problems of timing caused by the division of labor. Conducted at a community-household level of analysis, Linares' work exposes the contradictions that can develop in the interactions between technological coping on the one hand and the dynamics of social organization on the other. It shows that although, as an earlier review of domestic groups (Yanagisako 1979) asserts, similar production systems can be carried out by a variety of household organizations, *not all such organizations are equally effective.*

Linares' research is characterized by an instructive methodological rigor – a controlled three-village sample, the compilation of substantial quantitative data, and a research period spanning many years. The research demonstrates quantitatively and qualitatively the linkage between society and its resource base, such that the cropping system cannot be understood and analyzed apart from social organization. In this case, the key units for analysis are households, whose members are portrayed as struggling to mediate between the demands and timing of farming activities and the equally powerful demands and timing of social life, including the particular construction of the gender division of labor, inter-village networks of kinship relations and obligations, and the long-term cycling of ritual events. The latter, in this case, scarcely functions to regulate flows of nutrients in order to produce homeostasis (as an older ecology would have theorized), but rather creates a situation in which individual farmers must choose between coping socially and coping agriculturally. The choices the farmers must presently make, to their own detriment, provide the most compelling reasons that can be adduced for adopting the sociocentric or practice approach of this volume, and for further pursuing the research questions that an individual-based ecology of practice might raise. As Linares states:

> Individuals organized into broader relational spheres partake of common productive goals. They are part of what Shipton (1990:381–82) has called "cultural economy": the complex networks of social, economic, and political relationships that reflect, simultaneously, cultural values surrounding age, gender, and power relations and the more tangible demands of ecology, economy, and production.

Commenting on the earlier oral version of Linares' paper, Professor R. McC. Netting describes it as providing a

> down-to-earth demonstration that social organization is not some unmediated *Ding-an-Sich* as Yanagisako [1979] would have us believe. Under conditions of ecological uncertainty, some divisions of labor *do* work better than others. (Netting 1991:6)

He further goes on to state:

> The Jola cases give us a cross-section of farming systems ranging from Sambujat's intensively tilled, fertilized, transplanted, ponded paddies to the upland millet and groundnuts of Fatiya's plowed and presumably fallowed fields. The parallel social spectrum is even more fascinating with small monogamous, non-extended, non-ranked households cultivating with complementary male and female skilled labor at the intensive end, contrasted to large, extended, internally and externally ranked households with crops and tasks rigidly distinguished by gender. Of course, neither the agriculture nor the social organization falls into such neat dichotomies, but straddling the fence with a mixed strategy turns out to be dangerous. We learn that hapless Jipalom is moving to the Fatiya pattern of growing more millet to eat and letting women do all

the rice work, while men raise more groundnuts to sell. (Netting 1991:6–7)

Netting discusses other alternatives, including the possiblity of (and necessary conditions for) Jipalom society evolving in the direction of Sambujat-style labor and gender relations. His commentary focuses, then, on the alternative labor arrangements that work well in the timely mobilization of a labor force in particular contexts in comparison to those that do not, thereby emphasizing the value of Linares' meticulous presentation of agricultural production in villages with profoundly different labor arrangements. My own concerns from the perspective of the ecology of practice are with the social-system-maintaining, but agriculturally self-defeating, choices that are consistently made by the people of "hapless Jipalom." I wonder about the inevitability of its gender-based labor arrangements "moving to the Fatiya pattern" and conclude that it is precisely in the area of the relative organizational flexibility of communities, or the lack thereof, that some of the most important future discussions in the ecology of practice will likely take place.

Haratine and Bidan in the Senegal River Basin, Mauritania

The chapter "*Indirass* and the Political Ecology of Flood Recession Agriculture" by Thomas Park deals with the Mauritanian-side populations on the banks of the Senegal River. This is the northernmost (and lowest rainfall) of the research venues discussed in this book. Access to the river, a locus of water in a Sahelian dryland, is of great strategic importance to the farmers inhabiting its banks. The region is characterized by flood recession cultivation, and unpredictable flooding is an important ecological consideration affecting the availability of productive land. In this context, communal land tenure acts as a risk minimization strategy by redistributing the arable land annually to members of the community according to a specific order of priorities. Historically, several conflicting claims to land on the river's north (i.e., Mauritanian) bank by different ethnic groups (Bidan, Haratine, Halpulaar-en) have arisen. Each claim could be considered legitimate, and the local conflict has been intensified by interest in the area generated by a recent drought and major infrastructural development (i.e., the construction of the Manantali Dam and associated irrigation perimeters).

Park intimates that "wiping the slate clean" and beginning anew could be the best possible solution to current development dilemmas;

however, this leads to a further problem of the kind of tenure to put in place in reconceptualizing the relationship between flood recession farming and the larger sociocultural system. As an alternative to Western, neoclassical economic approaches to land tenure systems, based as they are on individual ownership documented by written deeds, Park proposes one that is based on Islamic and customary principles and grounded in the local ecology. In particular, this is a tenure system predicated on the Islamic legal concept of *indirass*, which refers to the obliteration of all traces of cultivation or habitation by the passage of time, along with a legal tradition in which current claims via oral testimony take precedence over those of the past. Park also advocates a traditional system of low-input flood recession agriculture geared to local consumption, over high-input irrigation agriculture geared to export markets as commonly advocated by Western development policies.

Two case studies from the Senegal River Basin illustrate the ways in which local communities have approached the legal principle of *indirass* in the context of different ecological, demographic, and political conditions. In them, Park demonstrates clearly his thesis that "political and ecological considerations enter into the application of Islamic principles." On this basis, he argues that a land-tenure system built upon oral testimony and current practice is more appropriate for development efforts in societies where literacy and careful written records are unlikely to be commonplace for some time. He finds this approach advantageous in terms of the flexibility it gives to the system in responding to changed conditions and ecological risks, while retaining the ability of the local people to understand and control the system.

Concerning this paper, Netting comments:

> Tad Park has done us all a service in exploring the functions of Islamic ownership rights that lapse with non-use of land and are legitimized by the oral testimony of the community. Written land registries and the granting of property rights by arbitrary state action led immediately to expropriation and population expulsion by Bidan. By comparison the horrors of the enclosure acts in England were a picnic in the park.... The distinctive common property portfolios that allowed groups to reallocate land consonant with the unpredictable levels of the flood were a splendid example of institutional adaptation but one literally not imaginable under the rubric of neo-classical economics. What might be dismissed as quaint (and inefficient) primitive communism was in fact complex, flexible, and embodying a clear hierarchy that prioritized

rights to cultivate. The investment that accompanies intensive agriculture is unnecessary in shifting cultivation and impossible in a flood
recession system where land is not a scarce good. Add to that a market
even *more* imperfect than the one under Western capitalism, and the
Mauritanian legislation [establishing privatization] as supported by
military coercion and the courts becomes not ignorant and misguided
but a sinister conspiracy. (Netting 1991:4–5)

In this case, cultural and political ecology and the ecology of practice
agree in extolling the virtues of preserving intact the system Park describes. Yet the ecology of practice would look further, not just at the
annual portfolio of available land and its potential uses, but also at the
population of hierarchically differentiated individuals whose access to
livelihood is successfully being "prioritized" by religio-juridical principles. It is the system of hierarchical ordering of persons that is perhaps
troubling and in need of additional analysis and clarification, not (any
longer) the principle of tenure that reassigns land to people when each
year it disappears under floodwaters and reappears transformed.

Haalpulaar in the Senegal River Valley, Senegal

The chapter by John Magistro, "The Ecology of Food Security in the
Northern Senegal Wetlands," takes us to the other side of the Senegal
River Valley for a multiple-village comparison of agricultural practice
and change. Magistro's chapter addresses the local-level aspects of
dam-based development replacing recession cultivation in the Senegal
Basin. Concerning the larger geo-political context of this development,
Netting commented:[3]

> The Senegal Valley scenario reminds us, if we ever forgot, that colossal
> meddling at national and international levels can be a tragedy that
> sacrifices human life and livelihood, wreaks environmental ruin, and
> destroys time-tested social institutions. The Manantali Dam [on the
> Senegal River] with its sacred, globally unquestioned modernizing
> premises of technological nirvana through electric power, irrigated
> crops, and river navigation is world-class hubris. International funders,
> engineers, and the African nations and consortia that share the addic-
> tion to gigantomania simply cannot imagine a flood recession system
> where temporary fields of sorghum, nomadic pastoral herds, mobile
> fisher folk, and opportunistic wood cutting ebb and flow in orderly and
> efficient fashion across a naturally chaotic environment.
>   The recognition of development insanity is not new – French ir-
> rigated perimeters, sometimes with advice from Chinese wet rice
> specialists, and dependent on diesel pumps, bulldozer levelling, concrete

ditches, improved seed, and chemical fertilizers have been proven ... to be unreclaimable economic disasters, and ... to introduce forced migration, seasonal stress, impoverishment, and crop appropriation at the muzzle of a gun. And anyone willing to count can discern that returns on labor and other inputs are so much higher under a flood recession regime than with the transplanting and other "fine-comb" techniques of Asian wet-rice cultivation that to adopt them where arable land is still abundant would be irrational if not demented. ...

Any TV news program on Africa still carries the overt message that it is either modern (i.e., American) agriculture and a capitalist market economy or famine for the continent. This devalues subsistence production, flexibility in the face of climatic variation, risk reduction, household labor and management, and local tenurial arrangements. The *hegemonic* ideal of large-scale, mechanized, energy-intensive, monocropped, market-oriented, industrial agriculture is enough to make *this* garden-variety materialist into not just a political economist but a raving symbolic anthropologist to boot, if you like, a kilo-Gramscian. (Netting 1991:3–4, 5–6)

Magistro's chapter rescues us from this "post-modern perdition" by presenting a fully contextualized, carefully quantified comparative study of the production regimes of several Senegal Basin villages. Here, the operative variable is the established pattern of recession cultivation and multiple-resource management practices versus a pattern likely to increase in the future in which farming options have been lost and only a core of reliance on rice irrigation perimeters remains, much to the detriment of the village producers.

More than this demonstration of the adaptiveness of indigenous management systems, however, Magistro's chapter shows the ecological significance in established societies of social differentiation and plural social groupings. The main study village population is divided into ten named strata, itself a sampling of the total of 24 named groupings in the valley. This variability, too, does not take into account the additional differentiations of age and gender that may further affect the status of persons whatever social stratum they occupy. Although comparative food security throughout the year for the village may depend in part on exploitation of numerous field types and resources, it also appears to rely on variability in human resources. At this stage, we can only note that the social variability attested to by Magistro is so salient – and parallel to the environmental and resource-use variability – that it requires explanation.

One potential explanation may be found in a recent discussion by Guyer (1993) and Guyer and Belinga (1995) expanding on the notion of wealth-in-people – a key "gatekeeper" concept of African ethnography focusing on the numbers of adherents of a group – to include the associated "wealth-in-knowledge" of these people as a construct particularly pertinent to Equatorial Africa. And although in the Equatorial context under investigation "devotion to knowledge went far beyond the basic requisites for making a living, even in a complex [rainforest] environment" (Guyer and Belinga 1995:93), the case that Magistro presents suggests that local social and cultural variability is as important in the Senegal River communities he studies as is the local variability in resource structure and management. Pursuing this question is evidently a significant direction for further research in the ecology of practice, which would seek to examine how individual, socially differentiated farmers might employ variable bodies of cultural knowledge in the context of local processes of agricultural production and social competition.

The Mende of the Gola Forest, Sierra Leone

In the chapter entitled "Shifting Social and Ecological Mosaics in Mende Forest Farming," Melissa Leach examines gender relations and changes in food crop production over the past 30 years in the rainforest environment of eastern Sierra Leone. In particular, she describes shifts in farming patterns among the Mende in this region in terms of the frequency with which and the way in which different types of farm sites are used. She argues that these changes are explained better as the dynamics of social and gender relations of labor and resource control in an evolving economic context than as the commonly cited effects of population pressure and environmental degradation.

Traditionally (before 1960), the Mende of the Gola forest chiefdoms relied on the production of rice and other crops using a system of shifting bush fallow cultivation within their forest environment. Since 1960, however, the Mende have increasingly been integrated into the larger economy as the production of commercial tree crops (cocoa, coffee, palm oil) has expanded and farmers' need for and use of cash to purchase goods and services has increased. This dynamic has resulted in complex changes in food crop production and control as mediated through gender and other social relations. Such changes include new production interests and opportunities, developments in farm organization and the size of the production unit, increasing concern about independent forms of production and food provisioning, new competing labor demands (i.e., between women and men, different crops, and household versus individual activities), and changes in extra-household labor arrangements. These organizational and economic changes result, finally, in subsequent changes in land use patterns in terms of choice of farm site type and the way in which each type is used. The most obvious general trends include changing patterns of crop production on the basis of age and status, whereby Mende women have become more responsible for household rice production as Mende men, particularly the elders, shift more into cash cropping, individual production, and smaller production units. Leach suggests that these changes in gendered food production further impact Mende gender concepts, and concludes that

> environmental management and changes, including those which ecologists might see as having negative or deleterious implications for the sustainability of agro-ecosystems, need to be understood in terms of different resource managers' interests and opportunities, and the dynamic relations between them.

In this case, Netting's comments suggested that his interpretations differed somewhat from those of the author, and his remarks merit quotation here at length.

> Unilinear is a word that anthropologists have not embraced since the days of Leslie White, and Melissa Leach is properly cognizant of just how complex ecological dynamics and social relationships are. Even the most prosaic patterns of agricultural site choice and cultivation are 'unevenly manifested in space, time, and across different social groups.' But despite our distrust of simple causes and our preference for history over evolution, there are some striking regularities in Mende movements from classic shifting cultivation toward incipient intensification. No one would deny the presence of increasing population pressure, scarcity of certain land types, and growing market involvement – this is West African déjà vu all over again – but the painstaking documentation of linked changes in work types, application, and amounts suggest commonalities of the division, multiplication, and gendering of labor that are also not unfamiliar. Though the details differ, what we see is intensification, using land longer in order to secure higher production per unit area and time, per hectare and per year. The Mende methods include shortening fallow proportionately by following rice with other crops, depending more on swamp plots where rice must be transplanted, intercropping with tubers, vegetables, and cotton, and emphasizing permanent kitchen gardening. Labor is steadier and less seasonally peaked in these pursuits. As I suspect is common among the swiddeners, these tasks are overwhelmingly female and result in an

overall *increase* in women's work. Ergo, woman the gatherer is also woman the intensifier. . . .

[Such] changes in productive conditions can bring the experience of material life into contradiction with those pervasive cultural representations of macho Mende males subduing the bush or women tending and domestically processing the crops. There *are* struggles over meaning, but these are accommodated within a broad and flexible set of concepts of people-bush relationships, gender interdependence and complementarity, and land tenure relationships which show no signs of fundamental change. "The dominant sense is of subtle shifts within an enduring repertoire." Less elegantly stated, behavior can and does change before meanings do. Ideology must play catch-up ball, unless of course the games of behavior and meaning are played out in the same league but in different parks.

But there we go, generalizing again, and glossing over the variety, the experimentation, the diversity of Mende economic and cultural creativity as they try on new farming methods without discarding the old cultural clothes that symbolize their identity and their past as a people. The danger of such adaptive modeling is that we ignore those situations when the state or those ethnic, class, or occupational groups with real power radically limit the choices and decisions that people make about their resources. (Netting 1991:1–3)

The problem that arises from Netting's analysis is to what extent "the state or those ethnic, class, or occupational groups with the real power" must exclusively be seen as external to Mende society. Are the Mende themselves change agents or only subject passively to adaptive processes? Are they free to engage merely in limited "cultural creativity" in terms of experimentation with farming methods, or do the subtle shifts within an enduring cultural repertoire described by Leach have real significance for the overall development and resource management ecology of Mende society? These are some of the questions that an ecology of practice orientation would pose and that would seem to be at the crux of the matter given the analysis of actual change that Leach provides.

The Susu of Kilimi, Sierra Leone

In "The Social Life of Swiddens: Juniors, Elders and the Ecology of Susu Upland Rice Farms," I examine patterns of farmer success and failure among Susu agriculturalists in northwestern Sierra Leone, focusing on the social and environmental contexts in which production problems occur. The Susu are small-scale agriculturalists who engage

in the swidden cultivation of upland rice and other grains and intercrops in a transitional Guinea savanna/moist deciduous forest environment. In this low-population-density "frontier" setting, labor rather
than land is the resource in limited supply, and highly seasonal rainfall makes labor mobilization a critical factor in successful agricultural
production. Moreover, Susu social organization is characterized by a
high degree of internal differentiation, most importantly on the basis
of age status and gender.

Analysis of individual farmer success, in terms of estimated farm
productivity (yields per hectare) and post-harvest stored yield, reveals
an overall deficit in food production and a substantial and patterned
variation in the ability of farmers both to produce a successful crop
and to maintain control of the crop at harvest time (i.e., store adequate supplies to acquire cash and stave off hunger). This variation is
explained in terms of the social system, including the social positions
of the farmers involved (elder men, junior men, and women), their
consequent ability to organize and mobilize labor, and the opportunity costs they undertake (e.g., the risks of intensifying production in
the face of substantial pressure on all farmers to redistribute crops and
labor). Thus, I argue in this chapter that the key problem in Susu
farming lies in the organization and mobilization of labor, which
causes management variations and sometimes farming failures, in part
because the crucial issue of farm site choice is often based as much on
social and organizational factors as on agro-ecological ones. In addition, I argue that the harsh environmental conditions faced by the
Susu, particularly the extreme rainfall seasonality of the Guinea
savanna environment, exacerbate production problems and food
deficits, thereby intensifying the pattern of social asymmetry and competition for labor among elder men.

The chapter concludes with a discussion of the general theoretical
implications of this case study, based on the analysis of the ways in
which the social and environmental conditions of Susu society together affect the choices made by individual farmers and the systemic
consequences of these choices. It is suggested that ecological anthropology needs a theory of resource management and exploitation
which incorporates processes of both coping (ecological adaptation)
and practice (action within a social context), in order to accomodate
the fact that

> the farmers' understanding of the environment is essentially a product
> of their social organization, as shaped by environmental constraints, in
> which different individuals have different statuses in society and express

these statuses through varying ways of competing for, using, and perceiving the crucial resources available to them.

In his comments, Netting indicated that this chapter

contradicts the ecological fallacy of Pollyanna functionalism, pointing out the Susu social asymmetries that lead to management failures.... You can't blame land, population, or technology when there is still an abundance of 30-year-old forest fallow. But socially the elders have the lion's share of everything – bigger fields, more rice, larger households, more polygynous marriages, more children, more dependents, and greater control of young men's work. Sounds like the [Gola] Mende [and] Fatiya ... cases perhaps? But the elder, like the big man Sahlins describes struggling up off the Chayanov slope, is ... driven to work harder himself, redistribute more, and manipulate the labor of others.... We need to know more about the social mechanisms of redistribution as elders compete politically in the "ecology of practice."

We might suspect that inequality and relations of exploitation may be *more pronounced* among shifting cultivators (including Park's flood recessionists) than among permanent intensive agriculturalists like the Sambujat Jola. Smallholder communities don't exhibit the ... redistribution of crops and labor that the Susu do, and property-holding households are not in a position to ... extort a surplus from one another, even when their heads differ in age, household size, and political clout, though they may all be subject to external elites. If and when Susu males ever have to take up the small plot swamp rice of their own women folk, elders will look less like witches and more like amiable husband-men. (Netting 1991:8–9)

Thus, the Susu case may exemplify the resource use practices of swiddeners in highly seasonal environments. Netting correctly identifies the crucial problem for future research in these societies, that of understanding the social relations (and agricultural implications) of redistribution of resources at harvest time. What is essential, however, is the demonstration in this chapter that the application of technology, farming success, and even the environmental impact of cultivation are variable outcomes of practice and the incorporation of resources (i.e., swidden sites) into individual social lives and contexts.

Rice Research in Rokupr, Sierra Leone

In "Toward an African Green Revolution? An Anthropology of Rice Research in Sierra Leone," Paul Richards investigates the actual conditions of rice research and the uptake of modern varieties in Sierra Leone in relation to the misrepresentation of this research and development change at the international level. In particular, Richards portrays the involvement of researchers as a significant variable in establishing the context of varietal availability in which local farmers make their crop selections. According to Richards, the dominant ideology guiding the spread of Green Revolution technology from Asia to Africa has been that high-yielding, fast-growing, semi-dwarf varieties (produced originally in Taiwan) provide the germplasm solution to the tropical world's rice-growing problems. This advocacy of a specific plant ideotype for all regions incorporates elements of religious-like belief and hegemony and the temporary suspension of any healthy skepticism or pluralism in science. Richards demonstrates his argument by a critical reading of a basic document in the history of international rice research (the proceedings of a conference on rice breeding, sponsored by the International Rice Research Institute at Los Baños, Philippines, in 1971), in which the leading proponents of the Green Revolution approach repeatedly emphasize the importance of faith in the majority opinion over heterodoxy and doubt in the applied sciences.

In contrast, the actual practice of rice research and development in Sierra Leone – as exemplified in the work of G. S. Banya and other national breeders at Rokupr Rice Research Station – has varied substantially from the orthodox model. Banya, for example, bred the rice varieties ROK 3 and ROK 16 from land races originally collected from his maternal uncle's farm in Kailahun in eastern Sierra Leone. As pure-bred strains these varieties were particularly suited to the local ecology, and recognized by local farmers as related to local breeds, but outperformed them by 10–20 %. Indeed, the actual uptake of modern varieties by farmers, as shown by Richards' 1987–1988 nationwide survey, has emphasized these familiar varieties (i.e., tall, tough wetland and upland rices) over the semi-dwarf "Green Revolution" types. Thus, despite the problematic orthodox breeder opinion promulgated by the Green Revolution, the impact of rice research in Sierra Leone has not been negligible. National breeders have found ways of supplying the needs of local farmers, and Richards advocates a new role for international crop breeders along the lines of the farmer-first work actually pursued at Rokupr.

The rationale for Richards' argument is bolstered by his survey of rice varietal uptake by a large sample of Sierra Leonean farmers. Undertaken in difficult research conditions, this survey demonstrates the extensive screening of rice varieties and experimentation with varietal choices by Sierra Leonean farmers. It further shows the general wide distribution of varieties throughout Sierra Leonean rural society, with apparently only the younger farmers having particularly limited access to modern varieties. Labor constraints apparently do not heavily affect varietal choice, although seedlessness may and is a key issue for rice breeding research and extension to address. Generally, farmers' decisions about varieties to plant are based upon pragmatic criteria, i.e., consideration of particular environmental factors that are apparent to the individual on the basis of past experience and present understanding. Richards' inference from this study is that "[e]mphasis must now be placed on a true anthropology of practice, that is, new rice types should be assessed from the perspective of *actual* performance in *real* (high stress) farming systems."

Finally, Richards discusses the future possibilities for rice research and development in the context of the current political economy of Sierra Leone, in which the local rice farming sector has habitually been ignored, imports from overseas of rice to urban areas have been emphasized, and now insurrection and coups d'état threaten to bring all regular research activity in the country to a complete halt. How should African crop researchers, expert at coping with sub-standard conditions, react to war and the challenge of post-war instability and recovery? According to Richards, technical options concerning low-input sustainable agriculture under unstable conditions would seem to have a renewed and radical relevance to debates about development policy and technology generation in the rice-dependent countries of western Africa.

CONCLUSIONS

The contributions to this book are focused on local issues in food crop production in western, sub-Saharan Africa, and because of the geographical focus the generality of the specific conclusions that can be drawn from the case studies is perhaps limited. But therein also lies the strength of these contributions. Each, by its focus on actual local practice, whether in the context of upland rice farming, recession cultivation systems, land tenure arrangements, labor mobilization, the gender division of labor, subsistence versus cash crop farming, or research and development in crop germplasm, has demonstrated the strength of a practice approach. The case studies particularly show that the greater the variability of the environment in which activity is occurring, the more applicable is the practice approach. The particular strength of this approach is therefore demonstrated in the very moment of admitting the difficulty in drawing general conclusions.

Some of the specific findings of the individual chapters deserve particular mention here. Linares' chapter, for example, instructs us in the variability of household structures and the problems created by organizational forms that, in some contexts, affect the mobilization and deployment of labor. She further raises the crucial question of the relative flexibility of social organization in relation to pressures of environment. This chapter and those by Leach and myself, which also focus on labor mobilization, should bring renewed attention to the labor process in ecology, perceived through the dual lenses of the ecology of practice, on the one hand, and cultural ecology, on the other.

Similarly, the chapters by Park and Magistro are significant in raising questions concerning social stratification in the chaotic environmental context that characterizes the flood recession regime of the Senegal River Basin. How do we understand stratification ecologically? Is stratification an institutional response to the environmental condition of resource variability, providing a solution to resource access problems via a legal principle that prioritizes the claims of resource users? Is it instead an adaptation to complex resource conditions producing, first, a valuable wealth-in-people needed to work the land and, second, a variable wealth-in-knowledge of how to exploit the resource base? The chapters provide some initial answers but raise further questions: what are the consequences for individual well-being of different statuses in the hierarchy? How mobile are individuals within the heirarchy? And how is the system ideologically maintained?

Leach's chapter on Mende farming in Sierra Leone supports the hypothesis that gender relations in agricultural production transform on their own dynamics. They are not direct functions of population pressure leading to intensification, as an argument developed on the premises of cultural ecology would have it. In the same vein, my study of Sierra Leonean Susu farming shows that hierarchy, and the incorporation of resources into social life via the hierarchy, results in the exacerbation of ecological problems. This work, explicitly focused on the ecology of practice approach, demonstrates that individual agency in society sometimes causes a failure of adaptation to occur. Finally, Richards' chapter shows that, in a larger context of inappropriate development packages, local scientists may be able to work out appropriate research practices. Richards emphasizes the essential rationality of local agricultural practice as exemplified by rice varietal choice in rural Sierra Leone. Furthermore, he raises the question of how, in the context of insurrection and a changing society, national rice researchers and local farmers could participate in training to exploit both indigenous and applied scientific knowledge concerning crop production.

Many questions remain, including in particular the practice implications of hierarchy, however flexible, in the river basins of the Mauritania and Senegal border zone and the swamps and uplands of northern and central Sierra Leone. Yet ultimately, the key implications of these case studies are particularly methodological and programatic. As guides to future work, they suggest that the most valuable research to be done in ecological anthropology now is particularistic and highly focused on individual productive activity in specific environments appropriately socioculturally contextualized. Ecological problems faced by local people, of course, may arise at institutional and environmental levels that, more often than not, extend beyond the local area and, therefore, warrant a political ecology framework for understanding and analysis. This is particularly the case in contemporary western Africa in terms of insurrection and warfare (e.g., Richards 1996). Nevertheless, it is the constellation of factors of local-level labor and social organizational arrangements that are crucial in determining how environmental problems, whatever their origin, will be perceived and responded to by individual actors. The ecology of practice, then, can be best understood as an approach and method for analyzing the local, social dimensions of global environmental change.

END NOTES

1. Similar views expressed in the literature include Karp and Maynard's (1983) re-reading of *The Nuer* and Collins' (1991) review of women and social reproduction in relation to sustainable development.

2. Compare this formulation, for example, to the Marxist notion that "production is a double nexus in which the relationship between humans and their environment is bound up with relationships among humans themselves (Marx 1913:273)" (cited in Collins 1992:179).

3. These comments were orginally directed to a paper by Michael M Horowitz and Muneera Salem-Murdock, entitled "Farming (and Fouling) the Floodplain," also on the topic of the Manantali Dam and its impact on local populations. Netting's remarks in this case apply with equal force to Magistro's paper, written later and directly substituted for the Horowitz and Salem-Murdock contribution.

REFERENCES

Appadurai, Arjun
    1986   Introduction: Commodities and the Politics of Value. *In* The Social Life of Things: Commodities in Cultural Perspective. Arjun Appadurai, ed. Pp. 3–63. New York: Cambridge University.

Bailey, F. G.
    1969   Strategems and Spoils: A Social Anthropology of Politics. Pavilion Series in Social Anthropology. Oxford: Basil Blackwell.

Barlett, Peggy
    1980   Agricultural Decision-Making: Anthropological Contributions to Rural Development. New York: Academic.

Barth, Frederik
    1959   [1965] Political Leadership Among Swat Pathans. London School of Economics, Monographs in Social Anthropology, No. 19. London: Althone.
    1966   Models of Social Organization. London: Royal Anthropological Institute.
    1967   On the Study of Social Change. American Anthropologist 69: 661–669.
    1994   A Personal View of Present Tasks and Priorities in Cultural and Social Anthropology. *In* Assessing Cultural Anthropology. Robert Borofsky, ed. Pp. 349–361. New York: McGraw-Hill.

Bennett, John W.
    1976   The Ecological Transition: Cultural Anthropology and Human Adaptation. London: Pergamon.
    1993   Human Ecology as Human Behavior – Essays in Environmental and Developmental Anthropology. New Brunswick NJ: Transaction.

Blaikie, P., and H. Brookfield
    1987   Land Degradation and Society. London: Methuen.

Boserup, Ester
    1965   The Conditions of Agricultural Growth – The Economics of Agrarian Change under Population Pressure. New York: Aldine.

Botkin, Daniel B.
    1990   Discordant Harmonies: A New Ecology for the Twenty-First Century. New York: Oxford University.

Bourdieu, Pierre
    1977   [1972, orig. Fr. ed.] Outline of a Theory of Practice. Cambridge Studies in Social Anthropology, 16. Richard Nice, transl. Cambridge: Cambridge University.

Brooks, George E.
1993 Landlords and Strangers – Ecology, Society, and Trade in Western Africa, 1000–1630. Boulder CO: Westview.

Collins, Jane L.
1991 Women and the Environment: Social Reproduction and Sustainable Development. *In* The Women and International Development Annual, Volume 2. Rita S. Gallin and Anne Ferguson, eds. Pp. 33–57. Boulder CO: Westview.
1992 Marxism Confronts the Environment: Labor, Ecology and Environmental Change. *In* Understanding Economic Process. Monographs in Economic Anthropology, No. 10. Sutti Ortiz and Susan Lees, eds. Pp. 179–188. Lanham MD: University Press of America.

Conklin, Harold C.
1961 The Study of Shifting Cultivation. Current Anthropology 2: 27–61.

Di Castri, Francesco
1976 International, Interdisciplinary Research in Ecology: Some Problems of Organization and Execution. The Case of the Man and the Biosphere (MAB) Programme. Human Ecology 4(3): 235–246.

Durham, William H.
1995 Political Ecology and Environmental Destruction in Latin America. *In* The Social Causes of Environmental Destruction in Latin America. Michael Painter and William H. Durham, eds. Pp. 249–264. Ann Arbor: University of Michigan.

Firth, Raymond
1951 The Elements of Social Organization. Boston: Beacon.

Ferguson, James
1992 The Cultural Topography of Wealth: Commodity Paths and the Structure of Property in Rural Lesotho. American Anthropologist 94: 55–73.

Giddens, Anthony
1979 Central Problems in Social Theory: Action, Structure and Contradiction in Social Analysis. Berkeley: University of California.
1984 The Constitution of Society: Outline of the Theory of Structuration. Berkeley: University of California.

Greenberg, James B., and Thomas K. Park
1993 Political Ecology. Journal of Political Ecology 1: 1–12.

Guyer, Jane I.
1981 Household and Community in African Studies. African Studies Review 24(2/3): 87–137.
1993 Wealth in People and Self-Realization in Equatorial Africa. Man 28(2): 243–265.

Guyer, Jane I., and Samuel M. Eno Belinga
 1995   Wealth in People as Wealth in Knowledge: Accumulation and Com-
        position in Equatorial Africa. Journal of African History 36(1):
        91–121.
Harner, Michael
 1977   The Ecological Basis for Aztec Sacrifice. American Ethnologist 4(1):
        117–135.
Harris, Marvin
 1977   Cannibals and Kings: The Origins of Cultures. New York: Vintage
        (Random House).
 1979   Cultural Materialism – The Struggle for a Science of Culture. New
        York: Random House.

Horowitz, Michael M
 1991   Victims Upstream and Down. Journal of Refugee Studies 4(2):
        164–181.

Huss-Ashmore, Rebecca
 1989   Perspectives on the African Food Crisis. Introduction. *In* African
        Food Systems in Crisis, Pt. 1: Microperspectives. Rebecca Huss-Ash-
        more and Solomon H. Katz, eds. Pp. 3–42. New York: Gordon and
        Breach.

Janzen, Daniel H.
 1973   Tropical Agroecosystems. Science 182: 1212–1219.

Johnson, Allen
 1983   Machiguenga Gardens. *In* Adaptive Responses of Native Ama-
        zonians. Raymond B. Hames and William T. Vickers, eds. Pp.
        29–63. New York: Academic.

Karp, Ivan
 1986   Agency and Social Theory: A Review of Anthony Giddens. Ameri-
        can Ethnologist 13(1): 131–137.

Karp, Ivan, and Kent Maynard
 1983   Reading *The Nuer*. Current Anthropology 24(4): 481–492.

Kopytoff, Igor
 1986   The Cultural Biography of Things: Commoditization as Process. *In*
        The Social Life of Things: Commodities in Cultural Perspective. Arjun
        Appadurai, ed. Pp. 64–91. Cambridge: Cambridge University.
 1987   The African Frontier: The Reproduction of Traditional African So-
        cieties. Bloomington: Indiana University.

Kopytoff, Igor, and Susanne Miers
 1977   Introduction – African Slavery as an Institution of Marginality. *In*
        Slavery in Africa: Historical and Anthropological Perspectives.
        Suzanne Miers and Igor Kopytoff, eds. Pp. 3–81. Madison: Univer-
        sity of Wisconsin.

Lees, Susan H., and Daniel G. Bates
1990   The Ecology of Cumulative Change. *In* The Ecosystem Approach
       in Anthropology: From Concept to Practice. Emilio F. Moran, ed.
       Pp. 247–277. Ann Arbor: University of Michigan.

Lewellen, Ted C.
1992   (2nd edition) Political Anthropology: An Introduction. Westport
       CT: Bergin and Garvey.

Linares, Olga F.
1992   Power, Prayer and Production – The Jola of Casamance, Senegal.
       Cambridge Studies in Social and Cultural Anthropology, No. 82.
       Cambridge: Cambridge University.

Little, Peter D., and Michael M Horowitz, with A. Endre Nyerges, eds.
1987   Lands at Risk in the Third World: Local-Level Perspectives. Insti-
       tute for Development Anthropology, Monographs in Development
       Anthropology. Boulder CO: Westview.

McCay, Bonnie J.
1978   Systems Ecology, People Ecology, and the Anthropology of Fishing
       Communities. Human Ecology 6(4): 397–422.

Magistro, John
1993   Crossing Over: Ethnicity and Transboundary Conflict in the Senegal
       River Valley. Cahiers d'Etudes Africaines 130(33–2): 201–232.

Meillassoux, Claude
1981   [1975] Maidens, Meal and Money – Capitalism and the Domestic
       Economy. Cambridge: Cambridge University.

Malinowski, Bronislaw
1922   [1961] Argonauts of the Western Pacific. New York: Dutton.

Marx, Karl
1913   Contributions to a Critique of Political Economy. Chicago: N. I.
       Stone.

Murphy, Martin F., and Maxine L. Margolis
1995   Science, Materialism, and the Study of Culture. Gainesville: Univer-
       sity of Florida.

Netting, Robert McC.
1990   Population, Permanent Agriculture, and Polities: Unpacking the
       Evolutionary Portmanteau. *In* The Evolution of Political Systems:
       Sociopolitics in Small-Scale Sedentary Societies. Steadman Upham,
       ed. Pp. 21–61. New York: Cambridge University.
1991   Comments on the Session "The Ecology and Economics of Food
       Crop Production in Sub-Saharan West Africa." Annual Meeting of
       the American Anthropological Association, November 1991. Un-
       published ms. in possession of the editor.

1993   Smallholders, Householders – Farm Families and the Ecology of Intensive, Sustainable Agriculture. Stanford CA: Stanford University.

Netting, Robert McC., Richard R. Wilk, and Eric J. Arnould, eds.
1984   Households – Comparative and Historical Studies of the Domestic Group. Berkeley: University of California.

Nyerges, A. Endre
1987   The Development Potential of the Guinea Savanna: Social and Ecological Constraints in the West African "Middle Belt." *In* Lands at Risk in the Third World: Local-Level Perspectives. Peter D. Little and Michael M Horowitz, with A. Endre Nyerges, eds. Pp. 316–336. Institute for Development Anthropology, Monographs in Development Anthropology. Boulder CO: Westview.
1989   Coppice Swidden Fallows in Tropical Deciduous Forest: Biological, Technological, and Sociocultural Determinants of Secondary Forest Successions. Human Ecology 17(4): 379–400.
1992   The Ecology of Wealth-in-People: Agriculture, Settlement, and Society on the Perpetual Frontier. American Anthropologist 94(4): 860–881.
1996   Ethnography in the Reconstruction of African Land Use Histories: A Sierra Leone Example. Africa 66(1): 122–144.

Orlove, Benjamin S.
1980   Ecological Anthropology. Annual Review of Anthropology 9: 235–273.

Ortner, Sherry B.
1984   Theory in Anthropology Since the Sixties. Comparative Studies in Society and History 26: 126–166.

Ottenberg, Simon
1984   Two New Religions, One Analytical Frame. Cahiers d'Etudes Africaines 96(24–4): 437–454.

Painter, Michael, and William H. Durham, eds.
1995   The Social Causes of Environmental Destruction in Latin America. Ann Arbor: University of Michigan.

Palm, Risa I.
1990   Natural Hazards: An Integrative Framework for Research and Planning. Baltimore: Johns Hopkins University.

Park, Thomas K., ed.
1993   Risk and Tenure in Arid Lands – The Political Ecology of Development in the Senegal River Basin. Tucson: University of Arizona.

Patterson, Thomas C.
1987   Development, Ecology, and Marginal Utility in Anthropology. Dialectical Anthropology 12: 15–31.

Peet, Richard, and Michael Watts
  1993 Introduction: Development Theory and Environment in an Age of Market Triumphalism. Economic Geography 69(3): 227–253.

Piot, Charles D.
  1993 Secrecy, Ambiguity, and the Everyday in Kabre Culture. American Anthropologist 95(2): 353–370.

Rappaport, Roy A.
  1968 [1984] Pigs for the Ancestors: Ritual in the Ecology of a New Guinea People. New Haven: Yale University.
  1971 The Flow of Energy in an Agricultural Society. Scientific American 224: 116–133.
  1994 Humanity's Evolution and Anthropology's Future. In Assessing Cultural Anthropology. Robert Borofsky, ed. Pp. 153–167. New York: McGraw-Hill.

Richards, Paul
  1983 Ecological Change and the Politics of African Land Use. African Studies Review 26(2): 1–72.
  1985 Indigenous Agricultural Revolution – Ecology and Food Production in West Africa. London: Hutchinson.
  1986 Coping with Hunger: Hazard and Experiment in an African Rice-Farming System. London: Allen and Unwin.
  1992 Famine (and War) in Africa: What Do Anthropologists Have to Say? Anthropology Today 8(6): 3–5.
  1996 Fighting for the Rainforest: War, Youth, and Resources in Sierra Leone. London: James Currey for the International African Institute.

Shipton, Parker
  1990 African Famines and Food Security: Anthropological Perspectives. Annual Review of Anthropology 19: 353–394.

Smith, Eric Alden
  1991 Inujjuamiut Foraging Strategies: Evolutionary Ecology of an Arctic Hunting Economy. New York: Aldine de Gruyter.

Spooner, Brian
  1982 Ecology in Perspective: The Human Context of Ecological Research. International Social Science Journal 34(3): 395–410.
  1984 Ecology in Development: A Rationale for Three-Dimensional Policy. Tokyo: United Nations University.
  1987 Insiders and Outsiders in Baluchistan: Western and Indigenous Perspectives on Ecology and Development. In Lands at Risk in the Third World: Local-Level Perspectives. Peter D. Little and Michael M Horowitz, with A. Endre Nyerges, eds. Pp. 58–68. Institute for Development Anthropology, Monographs in Development Anthropology. Boulder CO: Westview.

Steward, Julian
  1955 Theory of Culture Change – The Methodology of Multilinear Evolu-
         tion. Urbana: University of Illinois.

Stonich, Susan C.
  1993 "I Am Destroying the Land!" The Political Ecology of Poverty and
         Environmental Destruction in Honduras. Boulder CO: Westview.

Swartz, Marc J., Victor W. Turner and Arthur Tuden, eds.
  1966 Political Anthropology. Chicago: Aldine.

Vayda, Andrew P.
  1983 Progressive Contextualization: Methods for Research in Human
         Ecology. Human Ecology 11(3): 265–281.
  1986 Holism and Individualism in Ecological Anthropology. Reviews in
         Anthropology 13: 295–313.
  1991 Review of: Discordant Harmonies: A New Ecology for the Twenty-
         First Century, by Daniel B. Botkin. Human Ecology 19(3): 423–427.
  1994 Actions, Variations, and Change: The Emerging Anti-Essentialist
         View in Anthropology. In Assessing Cultural Anthropology. Robert
         Borofsky, ed. Pp. 320–330. New York: McGraw-Hill.

Vayda, Andrew P., and Bonnie J. McCay
  1975 New Directions in Ecology and Ecological Anthropology. Annual
         Review of Anthropology 4: 293–306.

Vayda, Andrew P., and Roy A. Rappaport
  1968 Ecology, Cultural and Noncultural. Pp. 477–497. In Introduction to
         Cultural Anthropology. James A. Clifton, ed. Boston: Houghton
         Mifflin.

Watts, Michael
  1983 Silent Violence: Food, Famine and Peasantry in Northern Nigeria.
         Berkeley: University of California.
  1984 "Good Try, Mr. Paul": Populism and the Politics of African Land
         Use. African Studies Review 26(2): 73–83.

Winterhalder, Bruce, and Eric A. Smith, eds.
  1981 Hunter-Gatherer Foraging Strategies: Ethnographic and Archaeo-
         logical Analyses. Chicago: University of Chicago.

Wolf, Eric
  1972 Ownership and Political Ecology. Anthropological Quarterly 45(3):
         201–205.
  1990 Distinguished Lecture: Facing Power–Old Insights, New Questions.
         American Anthropologist 92(3): 586–596.

Yanagisako, Sylvia Junko
  1979 Family and Household: The Analysis of Domestic Groups. Annual
         Review of Anthropology 8: 161–205.

CHAPTER 2

Diminished Rains and Divided Tasks: Rice Growing in Three Jola Communities of Casamance, Senegal

Olga F. Linares
Smithsonian Tropical Research Institute
Balboa, Panama

In fact, in African environments, the crucial problem for farmers has always been the timely and efficient mobilization of labor during the agriculturally productive rainy season. (Nyerges 1988:87)

This chapter examines how the age and gender division of labor, and the composition and scheduling of seasonal work groups, relate to the success of crop production. Among the Jola of Casamance, Senegal, the ability of household, family, and community members to mobilize agricultural labor efficiently during optimal periods in the yearly cycle depends greatly upon the structures of cooperation that are in place. Although it has been asserted that "a particular productive enterprise can be accomplished by a variety of productive units. . . ." (Yanagisako 1979:175), it has *not* been demonstrated that all productive units are equally efficient in accomplishing set tasks.

The household productive unit plays a central role in the cultivation of subsistence and cash crops. Thus, "domestic groups provide a pivotal element for the organization of agricultural production" (Swindell 1985:33); "much economic behavior is affected by long-term commitments to structures of relationships in households" (Cheal 1989:11); agricultural production is one of the activities "that are consistently associated with the small, numerous, corporate social units observers generally agree on calling households" (Wilk and Netting 1984:5); "the 'household' is important for the study of gender relations" (Smetsers 1995:3). In focusing primarily upon the household unit, however, scholars have tended to ignore the important tasks performed by other kinds of work groups. Doubts are being expressed as to whether households should continue to be treated as the privileged loci of economic activity. Some authors even assert that the household is "a concept that seldom fits rural African homesteads or compounds well. . ." (Shipton 1990:356). Other analysts insist that: "In the sphere of food production, the relevant social unit is not the household, the small co-residential group that shares a hearth, but the family, the larger group defined by descent and enlarged by marriage" (Weismantel 1989:63).

Clearly, productive tasks must be analyzed, not only at the household level, but also at higher units of aggregation. Individuals organized into broader relational spheres partake of common productive goals. They are part of what Shipton (1990:381–82) has called "cultural economy": the complex networks of social, economic, and political relationships that reflect, simultaneously, cultural values surrounding age, gender, and power relations and the more tangible demands of ecology, economy, and production.

Regional differences in the social formation of Jola labor relations are dramatic and pervasive. The local cultural economies that specify who performs what basic agricultural task, when, and with whom, vary markedly among groups living next door to each other. In the Esudadu region, both genders, and all generations, cooperate fully in producing a single crop of rice from the beginning of the rains. In the Kalunay region, women and men, constituted into distinct age groups, work totally independently of each other on different cropping systems. This frees them to perform the necessary gender-specific tasks promptly and efficiently. In the Kajamutay region, men and women, elders and juniors help each other with the tasks demanded by two very different cropping systems. This last alternative runs the risk of aggravating scheduling conflicts.

Under conditions of climatic uncertainty and ecological stress, the timely organization and management of agricultural labor becomes crucial. The hypothesis explored here, and confirmed separately by Posner, Kamuanga, and Lo (1991), is that in years of rainfall deficits it is particularly important for people to get the seed into the ground on time, so that plants can use the available moisture optimally to grow and develop. But because cultivators cannot predict in advance how the rainy season will develop, they usually proceed in customary ways, by activating institutionalized relations of cooperation and accepted mechanisms of labor exchange rather than by experimenting with new forms of organization.

The occurrence of drought and food shortages allows us to examine the dynamics of social aggregations from the perspective of stress. In parts of the Sahel and sub-Saharan Africa, seventeen drought-ridden years that began in the late 1960s eventually were to bring famine and desolation to large segments of the population. It was "one of the worst and most prolonged drought episodes in recent times" (Glantz 1987:38). The impact of these events on local populations has been analyzed from various perspectives: from the effects of colonial intervention on local crop production (Vaughan 1987), from the operation of mechanisms for entitlement and adaptive behaviors (Mortimore 1989), from recurrent failures in the development process (Glantz 1987), from "skyrocketing population growth" coupled with falling per capita food production (Hyden, Kates and Turner II 1993:401), and from the perspective of poverty and powerlessness, of economics as values and beliefs (Shipton 1990).

Although the Jola were spared the more catastrophic aspects of poverty and famine, declining rainfall clearly had negative effects on rice production. These effects can be analyzed from the standpoint of local social organization; that is, from the perspective of what kinds of social aggregations work best during periods of diminished rainfall or drought. Drought is not an easy phenomenon to define. There is "no simple correlation between meteorological drought and declines in agricultural production" (Glantz 1987:39). Seasonal variability in the distribution of rainfall is crucial, as are other environmental side effects, such as salinization, a drop in the water table, weed incursion, etc. A reduction of, say, 25% in long-term rainfall (a meteorological drought) may have mild adverse effects when compared with an agricultural drought when moisture at the right moment in the seasonal cycle is insufficient for the growth of particular plants (Glantz 1987:45). In Lower Casamance, the 1970s and 1980s were years of agricultural drought (more on this point below). During these years, some Jola groups, but not others, experienced severe crop losses and consequent food shortages. The hardest hit Jola groups were those that failed to respond to the diminished amounts and poor distribution of the rains by the timely mobilization of labor during the planting phase.

The first part of this chapter provides a brief historical synopsis of Jola settlement in Lower Casamance, including colonial developments and relations with neighboring peoples. This section helps us to understand why present Jola productive practices are so differently organized over relatively short distances. The following section outlines salient features of Lower Casamance ecology and the crop associations that dominate. The next section is an attempt to summarize regional social formations, with their customary social divisions of agricultural labor. Comparisons are focused upon three distinct but closely spaced Jola communities. A discussion of how variability in rainfall regimes relates to rice production comes after, as a prelude to the discussion of how the three farming communities responded to a particular dry year (1981). Cropping success in each of the three communities was measured in terms of returns to labor. Special emphasis was placed on the community of Jipalom, whose members obtained unsatisfactory yields. Although grounded in physical realities, their poor performance illustrates how customary patterns of reciprocity and aggregation may be inadaptive at times of stress.

A HISTORICAL SKETCH: DEVELOPMENTS SOUTH AND NORTH OF THE CASAMANCE RIVER

At the time of first European contact during the early part of the fifteenth century, the Jola (then known as Fulup) were confined to a coastal strip running from the entrance of the Casamance River south to the Cacheu River in Guiné Bissau (Almada, in Brásio 1964). The community of Sambujat, one of the villages in the Esudadu area selected for study, is located within this homeland. As a consequence of population disruptions and displacements related to the slave trade, groups of Jola began to expand out of their coastal homeland during the later part of the fifteenth century. They moved north across the Casamance River to regions occupied by peoples known as Bañun

(Bañuñ, Bagnun, Bainouk, etc.), whom they decimated or assimilated. The Jola moved by land and by water, using the tidal creeks or *marigots*, penetrating far inland. Following the Diouloulou *marigot*, the Jola came upon the fertile lands north of what is now the town of Bignona. The community of Jipalom in the Kajamutay represents the northernmost point of this expansion. It also represents a natural frontier, where rainfed agriculture was feasible in drought years, before reaching the vast alluvial plain bordering the Gambia River.

The social division of agricultural labor practiced in the communities of Esudadu and Kajamutay resemble each other more closely than they resemble the social division of labor in the communities of the Kalunay. The reason the Kalunay is so different from the other two Jola regions is that the Jola living there have been profoundly influenced by the neighboring Manding or Malinke peoples.

Beginning some time before the thirteenth century, and lasting well into the fourteenth century, Manding religious clerics and traders, originating ultimately in the Mali empire, came west to settle along the upper reaches of the Gambia, Casamance, and Geba Rivers. They were directly responsible for introducing the Qadiriyya version of Islam to the Jola populations there.

Together with Islam, the Manding slowly imposed aspects of their own organization upon the Jola living in villages with which they came into prolonged contact. Salient features of Manding social and political life found today among the Kalunay Jola communities they influenced – including the Fatiya community to be described later on – are the institution of village founders who own the village land and lend it to more recent immigrants; the incidence of extended, patrilineal households whose male members work under the leadership of the ranking household head; and a clear-cut gender division of labor. "For at least 300 years, rice cultivation among the Gambian Manding has been in the hands of women.... Millet and sorghum cultivation have been in the hands of men, at one point mostly slaves, for as long if not longer" (Linares 1992:187). The Manding division of labor is a result of their past history, when men served as warriors, merchants, and proselytizers living away from home. Nowadays, the Manding pattern of labor relations characterizes work in Kalunay communities such as Fatiya.

In addition to Islam and the aforementioned social institutions, the Manding were also largely responsible for introducing groundnut cultivation (i.e., growing peanuts) to the Jola living north of the Casamance River. Groundnuts were first cultivated in the 1850s by

Manding occupying the Middle Casamance, from where they slowly
spread to Lower Casamance. By 1935, the north-shore Jola largely
depended upon groundnuts for their cash income. The non-Islamicized south-shore communities of the Esudadu region, including
Sambujat, continued to reject groundnut cultivation, depending instead upon palm wine collecting for their cash revenues. These developments have been discussed in greater detail elsewhere (Linares
1992). Suffice it here to emphasize that the marked differences in crop
preferences, cultivation practices, and social procedures that we find
nowadays among the closely related Jola groups of Lower Casamance
are traced to the conflicts and contradictions created by conquest,
conflict, and cohabitation with culturally very different peoples.

GENERAL FEATURES OF POPULATION AND ECOLOGY

The Jola, who number some 375,000–400,000 persons, occupy the
southwestern corner of Senegal known as Lower Casamance.[1] Covering some 730,000 ha between 12–13° North latitude and 16–17° East
longitude, this territory stretches from the Atlantic Ocean to the Soun-
grougrou River, along both shores of the Casamance River
(Figure 1). Geopolitically, Lower Casamance is awkwardly wedged between the English-speaking country of The Gambia to the north, and
the Portuguese-creole speaking country of Guiné Bissau to the south.

The Casamance River flows gently east to west for approximately
300 kilometers. "About 40% of the area (300,000 ha) is comprised of a
complex of stream beds, inland valleys, and intertwining tributaries of
the Casamance River" (Posner, Kamuanga, and Lo 1991:1). These
tributaries form a dense network of *marigots* that provide the inhabitants with waterways for communication, fertile fishing grounds and,
before the drought, rice fields that were carved out from the mangrove
vegetation and desalinated over several years by flushing out the salts
with rainwater.

Low alluvial valleys or fans, where rainwater accumulates, are
everywhere topographically distinguishable from flat, sandy plateaus
no more than 30 meters in elevation, which are covered with derived
savanna and secondary brush vegetation. In protected areas a semi-
dry forest of the sub-Guinean type flourishes. This forest is composed
of tall trees of great widths (including *Khaya senegalensis*, *Parinari
excelsa*, *Erythrophleum guineense*, and *Ceiba pentandra*) and several
species of palm (the cultivated *Elaeis guineensis* and the *Borassus*
palm). Along the *marigots*, two species of mangrove vegetation (*Rhi-*

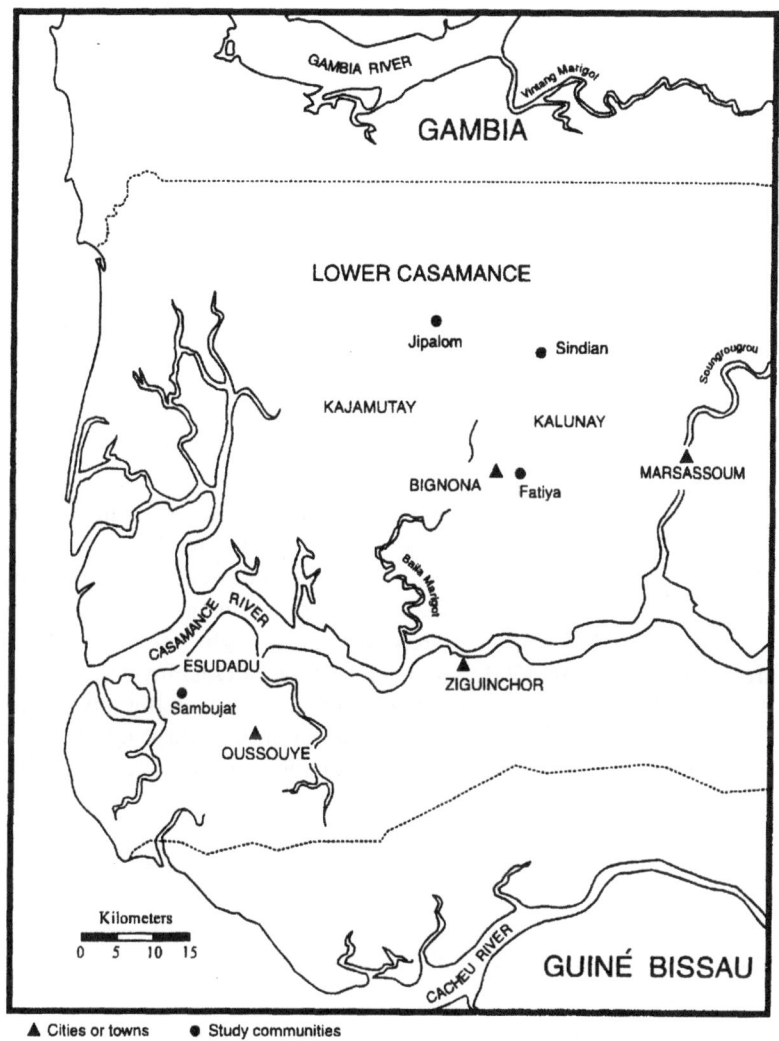

Figure 1. Map of Lower Casamance, Senegal.

zophora racemosa and *Avicennia nitida*) grow in the brackish waters that reach inland as far as 300 kilometers from the delta of the Casamance River.

In the alluvial depressions, or in lands adjacent to the *marigots*, all Jola regardless of location grow rice by direct seeding and transplanting in annually cropped, permanently cultivated fields. Everywhere, the Jola till the rice fields manually, either broadcasting the seed in upland fields or transplanting the seedlings onto flooded fields, and harvest the grain with the aid of a pocket knife. On the slightly raised areas, all Jola construct their villages, pasture their cattle, and protect their sacred forests.

Beyond these common procedures, however, agricultural practices vary markedly among various Jola sub-groups occupying different regions. North of the Casamance River, in the plateau communities of Jipalom and Fatiya, the Jola grow other crops in addition to rice, namely millet and sorghum for subsistence and groundnuts for the export market. South of the river, in the community of Sambujat, the Jola cultivate only rice, while obtaining cash from the seasonal collection of palm wine that they sell locally, for use in ceremonies to propitiate the spirit-shrines, or to outsiders such as Catholics living in the towns and cities.

Jola agriculture is heavily dependent upon a highly seasonal rainfall that is now concentrated in a 5-month period extending from June through October, with traces of precipitation occurring before and after. Generally speaking, rainfall decreases as one goes north. In the decades of the 1930s–1960s, preceding the drought, rainfall averaged from a high of 1600–1700 mm at the latitude of Oussouye, to a low of 1200–1300 mm at the latitude of Sindian, north of Bignona. Since these were "normal" years, having three months with a minimum of 250 mm of rainfall, rice agriculture flourished, and the granaries were full. Since 1968, however, declining precipitation, interspersed with years of severe drought, has had a disastrous effect on rice production. The physical dimensions of this phenomenon will be discussed below, but it is important to point out here that the drought had very different effects on different Jola groups depending upon the social dynamics which shape their agricultural practices.

JOLA SOCIAL ORGANIZATION: COMMONALITIES AND REGIONAL DIFFERENCES

The Jola are a coterie of agricultural peoples living in scattered, fairly autonomous communities, ranging in size from a few hundred to a few thousand inhabitants. Within any one village, agnatically related men and their in-married wives live in households that are arranged into courtyard groups (sg. *fank/hank*, pl. *unk*). These are clustered in turn into *quartiers* or wards, enjoying a large measure of independence from other like units. Social differentiation within and beyond the household is based largely on age and gender distinctions, although elements of ranking are present in the Kalunay. Politically decentralized or acephalous, all Jola place control over internal village affairs in the hands of household heads and elders residing within the wards. Chiefs, who were originally created by the French administration to facilitate their control over the population, are nowadays in charge of implementing the laws of the Senegalese State. They oversee tax collection, enforce school attendance, and supervise the work of government extension agents. Although chiefs everywhere are elected, in the Kalunay region the incumbent must be chosen from among members of the founding families.

Beyond these general features of Jola society, there is a great deal of sub-group, regional variation. Within less than a 50 km radius from Ziguinchor, the capital of Lower Casamance, social, religious, and economic patterns, including the organization of agrarian pursuits, differ radically. These differences can be summarized briefly with reference to agricultural practices in three communities: Sambujat in the Esudadu region to the south, Fatiya in the Kalunay region in the middle, and Jipalom in the Kajamutay area to the north. These communities are representative of distinct social formations, the result, in turn, of disparate social and political forces acting upon populations with slightly different historical backgrounds.

Although not chosen with fixed preconceptions in mind, or according to a strict random sample, these three communities were singled out for study because my own previous observations, and the subsequently published literature (Pélissier 1966), clearly indicated that they differed markedly in the social organization of agricultural work. At the time, I did not suspect that they also differed profoundly along many other social and political dimensions as well. All three villages were accessible by car, were relatively near each other, spoke mutually understandable subdialects, and were of generally equal (small) size. Characteristics that differentiate these communities, and the sub-regions they represent, include religion, marriage practices and household composition, cropping systems, land tenure, the gender and age division of agricultural labor, kin-based patterns of cooperation, work groups and hired hands, and the dynamics of emigration and wage labor.

Sambujat

With a population of just over 400 persons, this is the smallest of five Jola villages located in the non-Islamicized Jola sub-region known as

Esudadu to the south of the Casamance River, within the Department
of Oussouye. The majority of Sambujat's permanent residents still
practice the "traditional," pluralistic *awasena* religion (from the verb
kawasen, to pour libations of palm wine at the spirit-shrines). The
younger population that migrates seasonally to Dakar, however, has
become practicing Catholics. Made up entirely of locally born Jola,
the community is strongly corporate. Marriage is strictly monog-
amous, and village endogamy is the rule: 42% of marriages take place
within community boundaries and 48% take place with the community next door, a scant two kilometers away. Rice land is owned by
the male head of the conjugal family, who transfers a number of rice
paddies to his son at the time of marriage. The son, in turn, allocates
the best rice fields to his wife. Parents and their married sons continue
to cultivate in each other's rice fields. In the vast lowlands covering
67% of the area, the Esudadu Jola grow transplanted rice strictly as a
monoculture. Much of the work in the rice fields is done by the
conjugal family household, which tends to be small and non-extended
(Plate 1a). Within the household, each gender performs its own task

Plate 1a. Sambujat Farming: Two closely related couples cultivate together
in Sambujat. The men are tilling a paddy field using the flatter version of the
kajandu; the women are transplanting. (Date: 1981 agricultural season.)

sequences. The husband and his unmarried sons build up the bunds (embankments), ridge and furrow the bunded fields using two versions of a long-handled fulcrum-shovel (the *kajandu*), safeguard the maturing crop from bird destruction, and tie up the rice bundles after the harvest. The wife and her unmarried daughters carry fertilizer to the fields and also do most of the transplanting and harvesting. However, if time runs short, men may help with transplanting and harvesting. Though clearly drawn, the gender division of labor among the Esudadu Jola of Sambujat is not rigidly enforced.

Kin-based cooperation beyond the conjugal unit is also common: there is a great deal of inter-household cooperation. Two or more household heads, persons who belong to the same family even though they do not co-reside, often cultivate together on a regular basis. Thus, in one year, the units cultivating independently included 7 solitary men (widowers) and 27 conjugal households. In the same year, however, 33 other households cultivated in various arrangements: multiple households of father and married son, fraternal households of two married brothers, and grand households of father-son, plus two married brothers (see Linares 1984:422, Table 16.4).

In addition, group labor, in the form of large unisexual work groups, is constantly being engaged by Sambujat individuals in need of extra hands. These are organized work parties who hire themselves out for cash (see Swindell 1985:146). The general term for a work group is *embotai*, and each group is drawn from the members of particular residential units (Plate 1b). The largest of the male *embotai* includes all men and boys above eight years of age living in Sambujat. The next largest work group includes all the male residents of one of the two wards in which the village is divided. The smallest work groups include only the male members of two or more courtyard groups (sg. *hank*, pl. *unk*) within a single ward. In addition, all the young, unmarried boys in the community have their own work groups. Work groups are usually paid in cash, by the man or woman who contracts them. The money earned collectively is spent collectively, to buy sacrificial animals, palm wine, and condiments with which to hold particular ceremonies at one of the important village shrines. In addition to this cash, women and men have other sources of revenue. The principal source of cash for unmarried and married men comes from the dry-season trade in palm wine (Linares 1993). In this endeavor they earn as much money as other Jola earn in the groundnut trade. The main source of cash for women is the sale of small quantities of milled rice, baskets, and palm oil.

Plate 1b. A Sambujat men's work group (*embotai*) ridging and furrowing a paddy field belonging to one of its members. The field will be transplanted by the man's wife, with the help of other women, on the same day or shortly after. (Date: 1981 agricultural season.)

Among the women, cooperative work groups are also very common. They are drawn on the basis of the courtyard group (the *hank*) into which a woman is married. Thus, all the in-married women of *hank* "a," for example, make up one work group, and so on for all courtyards. The courtyard groups within one of the two village wards may join forces with each other. Women also work for money, which

Plate 1c. Members of a Sambujat unmarried girls work group (*club*) are helped by the young men to harvest before the abundant rice crop lodges or is eaten by birds. (Date: 1981 agricultural season.)

is pooled and used for ceremonies connected with the female-controlled *Sihuñ* shrines. The unmarried girls also have a work group, which they call a *club* (Plate 1c). They spend the money they earn from *club* work promoting dances, skits, and plays within the *maison des jeunes* and buying the necessary sound-making equipment. During the off-season, many of the youth of Sambujat leave the village to go back to school or to search for salaried work in secondary cities and in Dakar. In 1990, 60% of the girls and 52% of the boys came back during the rainy season (the time of the school vacation) to help their parents cultivate.

Fatiya

The second community selected for study is located within the Jola area known as Kalunay, to the east of the town of Bignona, in a region that has been strongly influenced by Manding peoples living nearby. The inhabitants of Fatiya (who numbered 320 in 1981 and 396 in 1988) are devout Muslims; the men pray together in their local mosque (the *miserei*), their children receive Koranic instruction in the local school, most inhabitants observe Ramadan and the *zakat*, and they cultivate for the *marabout* or local religious cleric. Marriage in Fatiya is polygynous and predominantly exogamous: 93% of the wives were born outside the community, an average of 25 kilometers away. Households are extended: the sons of a Fatiya man continue to live with their father even after they marry, enlarging sections of the parental home, unlike the sons of Sambujat residents, who construct their own houses when they marry.

As among their Manding neighbors, Fatiya men grow plateau crops, namely sorghum and millet for food and groundnuts as a cash crop, and women grow rice for family consumption in the alluvial depressions or fans. Men rarely visit the rice fields; women rarely visit the millet or groundnut fields.[2]

In contrast to the two other communities, the Fatiya population is divided between "owners of the land," who are descended from the village founders, and more recent immigrants. Members of the founding lineage own all the land. They lend plateau fields to the immigrant men and rice fields to immigrant men's wives in accordance with the needs generated by the developmental cycle of the particular immigrant's household. In order to continue borrowing parcels of land, or in order to be lent additional parcels, immigrants must demonstrate that they are using the land they have under cultivation properly and efficiently.

Married and unmarried brothers cultivate together under the leadership of the household head, who may be the father or, if he is resting, retired, or deceased, the oldest son. They use animal traction to plow the plateau fields, which are cultivated in millet, sorghum, and groundnuts in a short, bush-fallow rotational system (Plate 2a). Among the category of immigrant men, however, conjugal families, or at most two brothers who migrated together, do the cultivating. In

Plate 2a. Fatiya Farming: Two Fatiya brothers prepare their father's ground-nut and millet fields using the ox-plow. (Date: 1981 agricultural season.)

previous years, before animal traction was adopted, there was a great deal of cooperative, reciprocal labor exchange among the men. Nowadays, there is a great deal of reciprocal borrowing of oxen, plows, or carts.

Among the women, cooperative work in the rice fields is widespread and well-organized. Female work groups (called *sociétés*) are clearly structured. Again, using criteria from Swindell (1985:146), these groups can be described as "permanent or semi-permanent structures with elected officers, rules of procedure and fines for non-attendance." The elderly women, the married women, and the younger girls are constituted into their separate *sociétés* (Plate 2b). The first two groups do not regularly work for wages but rather engage in exchange labor. They work for each group member in strict rotation on the land that has been allocated to her by one of the village founders via her husband (Plate 2c). The girls, however, hire their labor out for cash.

All men, founders as well as immigrants who have borrowed land from the first, obtain cash from the cultivation of groundnuts. In addition to the *sociétés* work, women earn money by growing veg-

Plate 2b. A Fatiya married women's work group (*société*) prepares the rice field of one of its members. The tool they use is the *ebarai*, a long-handled hoe adopted from Manding women. Note that the women till the field flat, without making ridges. (Date: 1981 agricultural season.)

etables during the dry season in irrigated plots and selling them in the Bignona market. As elsewhere in Casamance, young men and young women are involved in seasonal migratory processes that increasingly tend to become permanent. According to M. Diop (1989:80), 18% of the men and 20% of the women, in five villages within the Department of Bignona, are seasonal migrants. In Fatiya, however, out-migration is fully counteracted by in-migration from new settlers, who come to settle in the village from land-hungry areas to the north.

Jipalom

The third community chosen for study (population about 600 persons in 1981 and 605 in 1988) is located in the Islamicized region known as the Kajamutay, to the north of the Casamance River. This is a slightly drier area, with more upland terrain and less lowland than the Esudadu. The majority of the Jipalom households are polygynous, with two or more wives present, and three or more generations also represented. Women marry out of their natal community but within

Plate 2c. A Fatiya married women's *société* harvestes the abundant rice crop produced during the 1981 agricultural season.

the local region; 83% of marriages take place within a circle of six villages, all within walking distance. Men enjoy important usufructuary rights to land from their uterine or matrilateral kin and also have rights to be protected and to obtain first fruits and chickens. Women continue to play an important role in their natal villages, where they take care of their brothers when they are ill and help them cultivate when they can.

The Jipalom Jola practice a dual agricultural system. In rainfed fields and, formerly, also in deeper parcels recovered from the mangrove by a process of flushing out the salt with rainwater, the Kujamaat grow rice for subsistence by using generally the same intensive techniques, including diking, bunding, and transplanting, as the inhabitants of the Esudadu region. In addition, however, they cultivate groundnuts for cash and some millet used for food using a short-fallow, rotational system whereby they cultivate the fields for two or three years and allow them to rest for five or more years.

All Jipalom male household heads own rice paddies and plateau fields. They are under the obligation to grant parcels to their sons when the latter marry, but whatever else they choose to do with their land, whether to cultivate it, lend it out, or leave it to rest temporarily, is up to each family head to decide. Although women do not own land, a widow or divorced woman can borrow rice fields from distant kin and affines.

Again, much of the agricultural labor is performed by the household. However, unlike the situation in Sambujat, and like the situation in Fatiya, the most common marital arrangement in Jipalom is polygynous. Within a single household two or more wives as well as three or more generations may be represented. Within one single ward, the majority of the Jipalom household members cultivate individually; only one multiple and two fraternal households cultivated together (Linares 1984:422, Table 16.4). In these "extended" households, a son will cultivate for his father, but not the reverse as in Sambujat.

In Jipalom, the gender groups and generations cooperate closely in agricultural work, as is also the case in Sambujat. The similarity in labor patterns between Jipalom and Sambujat has a basis in history (see above). However, in Jipalom as in Fatiya, but contrary to Sambujat, two different cropping systems have emerged following the introduction of groundnuts in the 1930s. Both women and men work on both kinds of cropping systems. Men and unmarried boys prepare the millet and groundnut fields while their wives or mothers help to seed and weed these crops and also help with the harvesting of millet and winnowing the groundnuts before bagging and selling (Plates 3a and 3b). Men ridge and furrow the rice fields, while their wives and unmarried daughters directly seed or transplant rice upon them, and also harvest the rice crop (Plates 3c and 3d). Because men and women must perform the same task sequences on two separate crops, the time spent preparing, planting, transplanting, or harvesting is inevitably drawn out.

In Jipalom, group labor is still practiced, especially in the rice fields. All the men of a *fank* or courtyard group, or all the men of a ward, prepare the fields together, using the *kajandu*. Elders usually work apart, slightly to one side. Similarly, all in-married women from the same *fank* regardless of age may transplant or harvest together. The young, unmarried girls, however, have their own work group. Like the work groups in Sambujat those of Jipalom are organized into work parties that hire themselves out for cash. They are paid a set amount, depending on the number of workers present.

Male household heads sell groundnuts for cash. With these earnings, or with cash from remittances sent by salaried sons living in town, they pay for extra labor. Women cannot sell rice, though they

Plate 3a. Jiplaom Farming: A Jipalom work group (*ekáf*) made up of the adult men from one ward ridges and furrows a groundnut field using the *kajandu*. The date is 1966. By the 1991 season, animal traction was beginning to be used (there were 18 ox-plow teams in the village), and cooperative labor was waning.

earn some extra money from pottery-making and processing salt. The amounts are small, however, which severely restricts the quantity of outside labor they can contract. As everywhere else, the Jipalom youth of both sexes are engaged in seasonal, migratory processes. They either go off to school, or they search for salaried employment in the cities. In recent years, about 24% of the boys and girls regularly come back during the rainy season to help their parents with agricultural work. This is less than half the percentage of Sambujat's young people who come back during the rainy season to help their parents with agricultural work. But it is higher than the 12.6% of young adults who had returned to Baland, a village in the Buluf area discussed by Lambert (1994:201), during the 1989 agricultural season.

In short, profound differences in the social organization of agricultural tasks characterize the three communities chosen for study. These differences are not simply structural; they have practical consequences for productivity, especially in dry years when conditions are less than optimal.

Plate 3b. An elderly woman and her daughter punch-hole seed the groundnut field prepared in 1966 by the men from one of the Jipalom wards. To this day, women help to seed the groundnut crop.

RAINFALL, DROUGHT AND RICE PRODUCTION

Rainfall in Lower Casamance varies significantly between regions. Figure 2 presents a comparison of precipitation in three rainfall stations located near the communities that I studied: the Oussouye station near Sambujat, the Bignona station near Fatiya, and the Sindian station near Jipalom. A fourth station (Ziguinchor), which has the longest rainfall record in the region, was included in the comparisons as a control. The 6-way comparison was based on randomized block analysis of variance (ANOVA). The results show that all combinations of sites, with the exception of Bignona and Sindian, are significantly different with respect to mean yearly rainfall in the years up to 1994.

Plate 3c. A Jipalom father and son prepare a paddy field using the *kajandu*. The date is 1981. Note how relatively dry the ground is.

Plate 3d. A group of in-married Jipalom women harvest a rice field during the 1981 agricultural season. Note the sparseness of the crop and the relatively slim pickings.

$N = 178$, $R^2 = 0.900$

variable	D.F.	F-ratio	P
site	3	19.656	0.000
year	73	12.069	0.000

Post-hoc Pair-Wise Multiple Comparison Fisher's Least-Significant-Difference Test Matrix of pairwise comparison

	Bignona	Oussouye	Sindian	Ziguinchor
Big.				
Ous.	0.000			
Sin.	0.168*	0.000		
Zig.	0.023	0.000	0.002	

* Bignona and Sindian are the only two sites that are not significantly different with respect to mean yearly rainfall.

Figure 2. A comparison of rainfall records in four Lower Casamance stations based on randomized block analysis of variance (ANOVA).

Variations in the social organization of rice-growing practices, however, do not strictly reflect variations in rainfall regimes. Thus, rice-growing procedures are very similar between Sambujat and Jipalom, which receive significantly different amounts of rain, but are very different between Jipalom and Fatiya, which receive similar amounts of rain. Clearly, it is Jipalom's rice-growing practices that seem to be out of phase. Like many other countries of Africa, Senegal has experienced a prolonged drought that began around 1967 and continued well into the 1980s. A 10-year running average of the total annual rainfall recorded in the four Casamance stations (Figure 3) shows that the area has become progressively drier during the last 30 years.

The monthly distribution of yearly rains is perhaps more important for the growth and development of the rice crop than inter-year variations. Again, the experts (Posner, Kamuanga, and Lo 1991:16) indicate that in upland areas, where rice is direct-seeded, an early beginning for the rains and a good distribution throughout the season are crucial. On the south bank, where transplanted rice predominates, abundant rains in the months of September and October are required for the plants to produce.

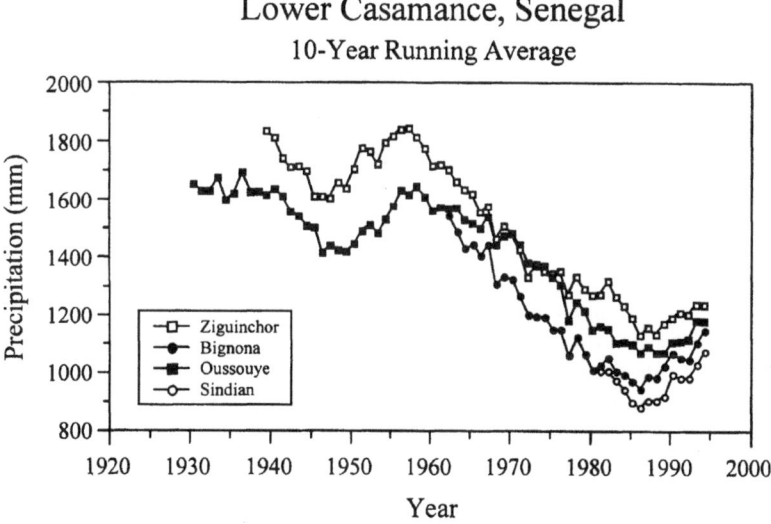

Figure 3. Graph representing 10-year running averages of precipitation in four Lower Casamance stations. Running averages, based on a statistical mean for several years, smooth out yearly variations, exposing long-term trends more clearly. Note that each point represents an average, not the particular rainfall for a specific year.

Beginning in 1968, however, the rainy season in Lower Casamance has actually become shorter. Taking Ziguinchor as an example (Figure 4), during the rice growing season (between May and November), the number of months having some precipitation (regardless of quantity) dropped from an average of 6.36 months in the years 1920 to 1967, to an average of 5.70 months in the years 1968 to 1994. These differences are statistically significant ($t = 2.749$, D.F. $= 26$, P $= 0.011$).

There is a strong correlation between rainfall and rice yields. As Posner, Kamuanga, and Lo (1991) point out, the combined effects of amount and distribution of rainfall, soil water reserves, and evaporation can bring drastic reductions in agricultural outputs. In Lower Casamance, during the 20-year drought period that concerns us here, about 50% of the rice-growing area was abandoned due to salt toxicity in the lowland fields adjacent to the *marigots* and drought stress in the upper fields further inland (Posner, Kamuanga, and Lo 1991:5). For the combined Departments of Bignona, Ziguinchor, and Oussouye, the total area that was cultivated in rice fell by half, from 40,908 ha in 1960–1964 to 20,854 ha in 1980–1984 (Posner, Kamuanga, and Lo 1991:9, Table 2). Almost the entire category of what were once highly productive mangrove paddies had to be abandoned.[3]

The effects of a dry year may be felt subsequently. Thus, as shown in Figure 5, 1973 was a year of low yields despite a rainfall of 1045 mm. This was because it followed two dry periods: 1971, with 905 mm of rainfall, and 1972, with 656 mm. By then, the ground had hardened, making it difficult to work the land, and people were out of seed. The same happened in 1978, which also followed a particularly dry year (1977, with 851 mm of rainfall).

Individuals can respond to diminished rainfall in several ways: they can broadcast the seed rather than transplant the seedlings; they can thin the nurseries and transplant only half the crop, leaving the rest in place; or conversely, they can fill in with plants from elsewhere. Farmers can also plant faster-growing, though lower-yielding, rice varieties, or they can concentrate on cultivating the *kuyelen*, the fields where rainwater accumulates, and ignore the *weng*, the normally deep-flooded mangrove fields. Finally, if they have the means, individuals can use associative labor and hire extra help to get tasks done on time. However, the efficacy of all these coping techniques depends upon some foreknowledge of how the rainy season will unfold. If, for example, the rains turn out to be abundant, women will be forced to transplant; they cannot direct sow seed onto flooded fields or the seed

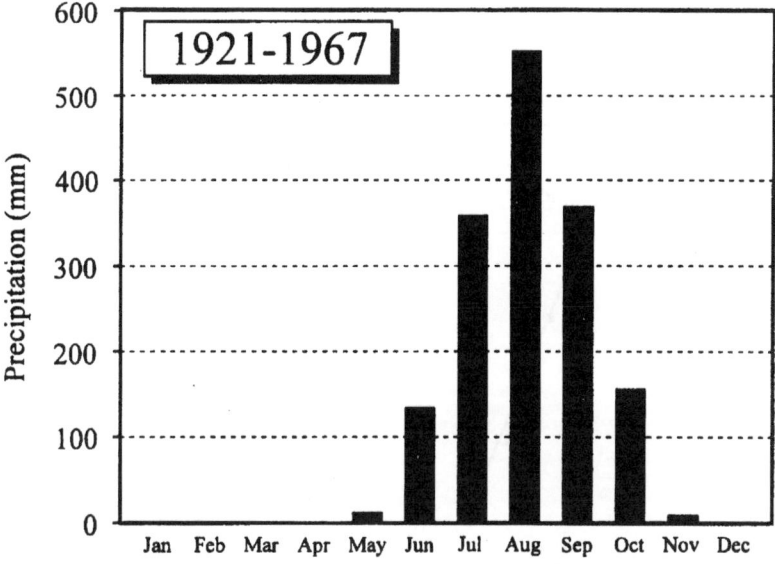

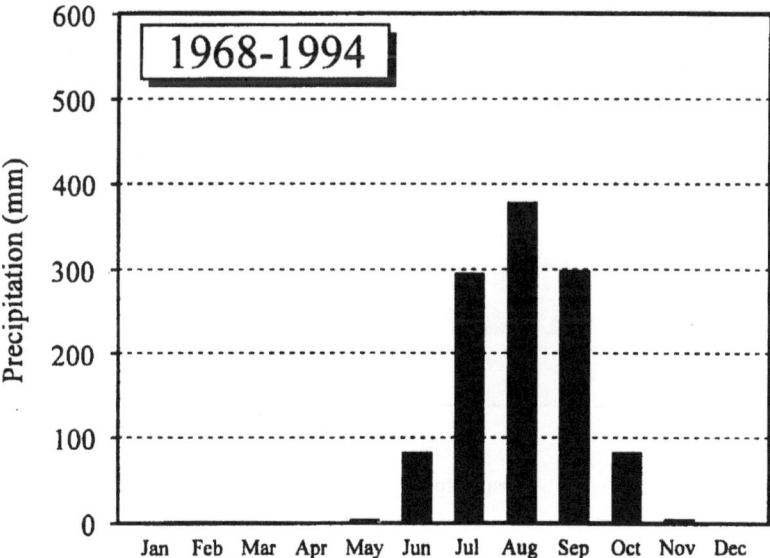

Figure 4. Months in the Ziguinchor station having some precipitation. Comparison of the graphs shows that the rainy season has become truncated following the drought that began in 1968.

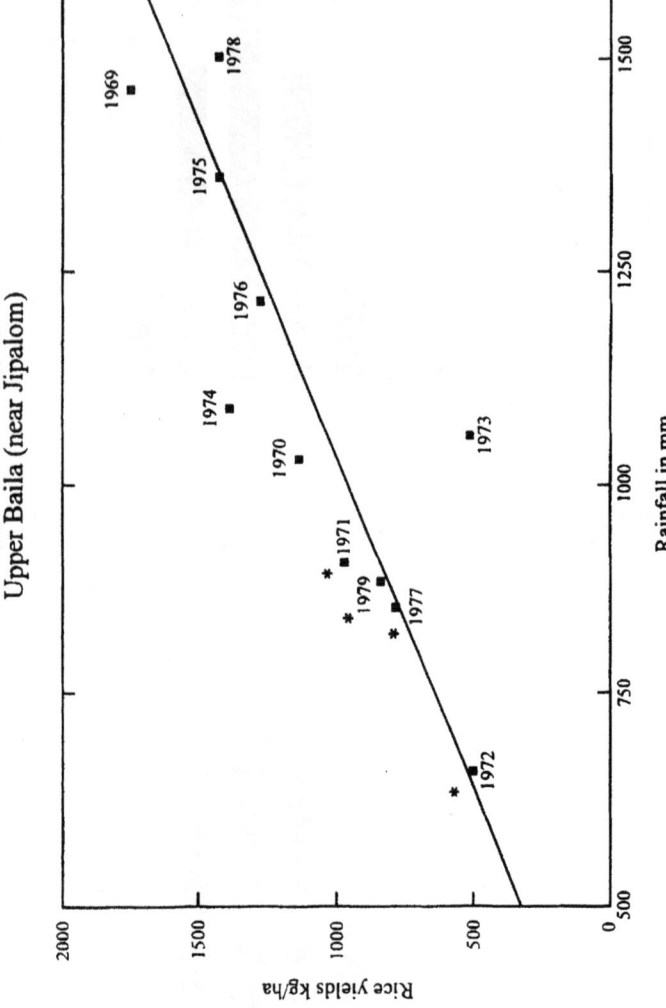

Figure 5. Rice yields are correlated with rainfall. Data from Louis Berger International (1981).

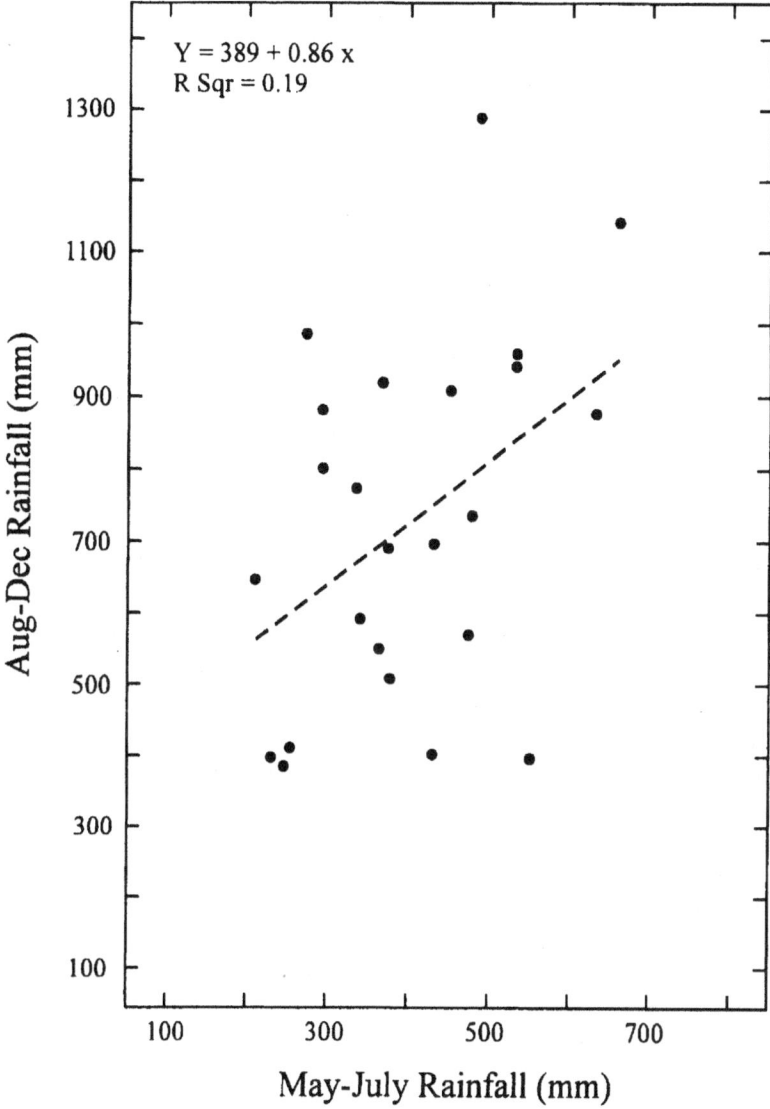

$Y = 389 + 0.86 \, x$
$R \, Sqr = 0.19$

Figure 6. A simple regression showing that there is no statistical correlation in rainfall amounts between the beginning and end of the rains.

will rot. Another simple regression (Figure 6) suggests that it is practically impossible to know from the first months of rainfall (May–July) what the last months (August–December) will bring. In other words, an r^2 of .19 indicates that rainfall during the first part of the season is only weakly positively correlated with rainfall in the second half of the season. As a result, farmers cannot safely predict what the last months will bring and are thus handicapped in making optimal decisions as to what to plant, when, and where. Under these circumstances, the social arrangements of production that are already in place may play a determining role in how well individuals cope with adverse ecological conditions. This proposition was tested with reference to the activities of farmers as they unfolded during one dry year in the three communities under study.

THE 1981 AGRICULTURAL SEASON

The year 1981 was not the driest of recent years in Lower Casamance, yet rainfall in all the stations was very near the minimum permissible level for rice agriculture. By November, it was clear that rainfall was about 200 mm below average in each of the sub-areas being considered. Although the study that I conducted on rice yields and returns to labor took place over a decade ago, the results still have important theoretical implications, showing that the manner in which labor in society is organized has immediate, practical consequences. Moreover, what happened in 1981 foreshadowed some of the changes that are now taking place.

Figure 7 summarizes the dynamics of task sequencing by gender during the 1981 agricultural season. What follows below is an account of how work in the rice fields unfolded in each of the three communities as the season progressed. The purpose of the discussion is to highlight disjunctions and discontinuities in the scheduling of agricultural work. Special attention is paid to when work was initiated, and by whom, for this largely determined absolute yields obtained and returns to labor.

In Sambujat, the more humid of the locations, conjugal family members began seeding in the rice nurseries promptly during the second week of July, finishing by the end of the month. It is essential to get the ground prepared while it is wet, so men began to build up the bunds, and ridged and furrowed the regular fields by the beginning of August. They tilled the deep, mangrove fields first to use rainwater optimally, then moved up the slope gradually to till the intermediate fields; the rainfed fields on the plateau were the last to be tilled and transplanted upon. During one day in early August, and in only one sector of the rice fields, I counted two men working alone, five men working with one or more sons, two sets of brothers and

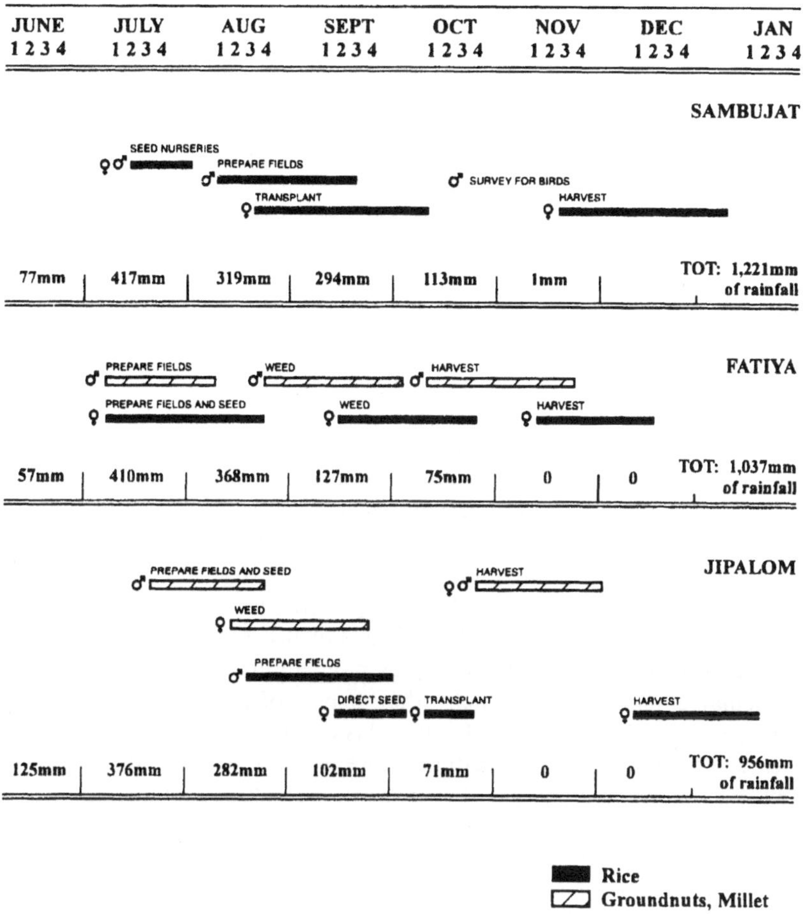

JUNE 1 2 3 4	JULY 1 2 3 4	AUG 1 2 3 4	SEPT 1 2 3 4	OCT 1 2 3 4	NOV 1 2 3 4	DEC 1 2 3 4	JAN 1 2 3 4

SAMBUJAT

SEED NURSERIES
♀♂ ▬▬▬▬ PREPARE FIELDS
♂ ▬▬▬▬▬▬
TRANSPLANT
♀ ▬▬▬▬▬▬
♂ SURVEY FOR BIRDS
HARVEST
♀ ▬▬▬▬▬▬

| 77mm | 417mm | 319mm | 294mm | 113mm | 1mm | | TOT: 1,221mm of rainfall |

FATIYA

♂ PREPARE FIELDS ▭▭▭▭   ♂ WEED ▭▭▭▭▭   ♂ HARVEST ▭▭▭▭▭
♀ PREPARE FIELDS AND SEED ▬▬▬▬   ♀ WEED ▬▬▬▬   ♀ HARVEST ▬▬▬▬

| 57mm | 410mm | 368mm | 127mm | 75mm | 0 | 0 | TOT: 1,037mm of rainfall |

JIPALOM

♂ PREPARE FIELDS AND SEED ▭▭▭▭   ♀♂ HARVEST ▭▭▭▭
♀ WEED ▭▭▭▭
♂ PREPARE FIELDS ▬▬▬▬
♀ DIRECT SEED ▬▬ TRANSPLANT ▬▬   ♀ HARVEST ▬▬▬▬

| 125mm | 376mm | 282mm | 102mm | 71mm | 0 | 0 | TOT: 956mm of rainfall |

▬▬ Rice
▭▭ Groundnuts, Millet

Figure 7. The 1981 season: dynamics of task sequencing by gender.

their sons working together, and a work group of 37 youths from the entire village working for one of the agnates. The women followed two weeks later, transplanting rice seedlings that had attained the right height onto the fields that the men had prepared. With the mangrove fields, the women must wait two weeks for the rainwater to dilute the salt, and for the latter to seep downward, before they transplant short-cycle varieties. In the intermediate and upper fields, however, the women must wait only two days after the fields are tilled in order to transplant. The plateau fields are clayey and take time to dry out, so they are transplanted with long-cycle varieties. Group labor was at its peak in September, when particular men or women hired organized work parties to complete land preparation or transplanting before the rains stopped in October. After a rest-period of about one month, during which time children guarded the crops against bird damage, and men went to their palm groves to tap palm wine, the rice harvest began in earnest in November. This is another time when demand for group labor is at its peak. The rice must be harvested before it lodges or is eaten by birds. Women who have the means to hire one of the work groups, or whose husbands have the means to help them pay, get their fields harvested first. By the beginning of January all fields were harvested and rice bundles, which were tied up by the men at the end of each harvest day, were left on the ground to dry before transporting them home to the granaries.

In Fatiya, both genders also initiated work in their respective crop systems early in the 1981 wet season–usually by the first week in July, a week earlier than in Sambujat. Men and women kept on with their independent tasks well into the second week in December. The men worked individually on their millet and groundnut fields, using animal traction. In 1981 only 11 household heads owned the complete equipment consisting of a pair of oxen, plus plow and cart. Men borrowed particular equipment from each other: a pair of oxen lent out for a day against the use of a plow for two days, a donkey one day against a cart the next day. The weeding, however, was done gradually by each man alone. Women also began work individually by hoeing one of their allotted rice fields first, then joining one of the organized women's work groups two or three times weekly to prepare each field in strict rotation. The technique they used was to break the ground using the long-handled female hoe (the *ebir*) then place the seed in straight, shallow furrows. Very few women transplanted. Despite considerable rain in the months of July and August, most Fatiya women continued to broadcast the seed of fast-growing varieties directly on the ground. They were right to do so because the fields promptly dried up during the following month. The switch from transplanting to direct seeding in rows, to facilitate weeding, had taken place during the 1970s as a response to the drought. By the end of November the men, and by the end of the third week in December the women, were through with their respective harvests.

People in the third community, Jipalom, began to cultivate late in the season. There was a delay of at least three weeks before men began plowing their groundnut and millet fields after mid-July. The immediate reason for this was not the rains; it had rained particularly well in June. The delay was partly because everyone had been involved in the circumcision rites taking place in the neighboring community of Jinunje. A community holds these important rites, without which the young men cannot marry, every 25 years or so. Although they are supposed to be held during the dry season, these ceremonies often drag on into the rainy season in order to permit those hosting the event to obtain enough food for feasting the guests properly. Many of Jipalom's wives and mothers came originally from Jinunje, and because they were female agnates of that community their sons had important matrilateral rights in the various Jinunje compounds, including the right to borrow extra rice fields when necessary. Thus, it was imperative that the men and women of Jipalom be present at the Jinunje festival, lest they forsake their own and their children's rights to borrow land and receive services from their uterine kin.

The delay in beginning work was further compounded by the gender division of cropping tasks. Men had to finish preparing their plateau fields before ridging and furrowing their wives' rice fields. Their wives were under the obligation to help them weed. By the third week in August, very few men had brandished their *kajandu* to till the nurseries or rain-fed fields. Not a woman was seen in the rice fields; they were all weeding their husbands' millet plots. As a result of these reciprocal obligations, women were both direct seeding and transplanting rice well into the end of October when the rains were already finished. Moreover, the mobilization of labor was not taking place promptly or efficiently but on an ad-hoc basis, and only among those individuals who could pay. The harvest did not start until December, but it did not last long. All in all, the pace of work had dragged behind the other two villages.

DROUGHT, YIELDS AND RETURNS TO LABOR: MECHANISMS AND IMPLICATIONS

The labor invested in rice-field cultivation varies greatly, depending upon the tasks to be performed. In flooded fields, transplanting is time-consuming. In direct-seeded fields it is the weeding that takes a great deal of time. Conversely, harvesting takes about the same amount of time whether the fields were broadcast-seeded or transplanted upon. All data concerning inputs and outputs are rough averages based on the amount of time it takes to perform various tasks.

Labor-Invested	N	Paddy/ha	Milled/ha[2]	N	Kg/day	R-L[3]	R-L[4]
			SAMBUJAT				
♀ 105 days/ha ♂ 108 days/ha	32			41			
213 days/ha		2252 kg	1464 kg		6.87	14.30	11.45
			FATIYA				
♀ 197 days/ha	18	2290 kg	1489 kg	28	7.56	15.75	12.60
			JIPALOM				
♀ 111 days/ha ♂ 76 days/ha	26			33			
187 days/ha[5]		1149 kg	747 kg		3.99	8.31	6.65

1. Return to labor is number of days worth of food earned by a day's labor.

$$\text{R-L} = \frac{\text{milled/ha}}{\text{days/ha}} = \frac{\text{kg/day produced}}{\text{kg/day consumed}}$$

2. Conversion from paddy to milled rice is 0.65.

3. The World Bank calculates that an adult consumes 176 kg milled rice a year or 0.48 kg daily.

4. The Jola themselves calculate that an adult consumes 220 kg milled rice a year or 0.60 kg daily.

5. This figure is corroborated by Posner, Kamuanga and Sall (1985:21).

Figure 8. Rice yields and returns to labor during the 1981 agricultural season.[1]

Figure 8 summarizes data gathered in 1981 on the time invested, and the amounts of paddy harvested, in the three communities under study.[4] This figure indicates that the amount of labor invested in rice agriculture in the three communities was roughly the same (i.e., 213 days/ha in Sambujat, 197 days/ha in Fatiya, and 187 days/ha in Jipalom). Yet the crucial figure of returns to labor (R-L) indicated in the last two columns, and defined as the number of days worth of food produced by a day's labor (given World Bank and Jola estimates for daily rice consumption), was not equivalent throughout the three locations. Despite a significant difference in rainfall, the R-L findings for Sambujat and Fatiya were quite comparable: about 12 days worth of food for one day of work in both locations. Thus, even though the technology of rice production and the social organization of task sequences differed markedly in these two communities, labor in both was mobilized rapidly and efficiently at the beginning of the rainy season. This was not the case in Jipalom, where returns to labor (R-L) were very much lower: about 6, or roughly half the number of days worth of food for a day's worth of labor obtained for Sambujat and Fatiya. Yet, rainfall differences between Jipalom (i.e., the Sindian station) and Fatiya (i.e., the Bignona station) are not significant.[5]

What are the implications of these results? Why did the people of Jipalom run into trouble? This study suggests that there are at least two effective options open to farmers in the face of rainfall deficits and a shortened growing season. From the first rains, both men and women can be out in the fields, as in Sambujat, cooperating in nursery preparation and seeding in production of the sole crop of rice. Or, conversely, gender groups can work autonomously, with men initiating soil preparation on their plateau fields of groundnut and millet, and women beginning to hoe their rice fields as soon as the first rains fall, as in Fatiya. A third option typified by the Jipalom gender division of labor, which is a historical offshoot of the Sambujat division of labor, turns out to be maladaptive in times of drought or reduced rainfall. A social system in which both genders cooperate in doing the agricultural work demanded by two separate cropping systems of rice and groundnuts/millet means that seeding and planting times are at a serious risk of being delayed and dragged out. In years when the rains are sufficient and well distributed, yields may not suffer. When I was first in Jipalom in 1964–66, the rains were extremely abundant, granaries were full, and both genders were involved in both cropping systems. In 1981, however, the same social division of labor proved disastrous.

The main reason for Jipalom's crop failure can be traced to the delay in planting time. Work by Posner, Kamuanga, and Lo has shown that there is a significant correlation between date of planting or transplanting and yields. As they state, "linear regression indicates that a delay of one week in the date of planting/transplanting results in yields lowered by approximately 55 kg/ha" (1991:47 and Table 23, pg. 49).

Even if the circumcision festival in the nearby town had NOT taken place, it is probable that yields in Jipalom would have suffered anyway. In the community of Suel, a few kilometers from Jipalom, "men first devote their energies to planting groundnuts and millet and do not till the rice fields until the end of July. By that point, they have little hope of obtaining good yields" (Posner, Kamuanga, and Lo 1991:55). On the other hand, I was able to confirm that the inhabitants of Balandine, a community so near Jipalom that its rice fields are contiguous, harvested a decent crop of rice in 1981 even though many persons also attended the Jinunje circumcision rite. The reason for this is that the people of Balandine, being Catholics, do not cultivate either millet, sorghum, or groundnuts, for these are considered to be crops grown by Muslims. Even though the men were late in starting, they were able to make up for lost time because they had only the rice fields to prepare.

CONCLUSIONS

One may ask why the Jipalom Jola have not adjusted better to recurrent droughts and arranged their agricultural practices accordingly. After all, there was a prolonged drought in the years 1909–1919 and several episodes of dry years in the period between 1938 and 1949. The answer, of course, is that people do not shed their social systems at will, or overnight. The particular way in which labor is structured and mobilized in a society is socially and culturally determined. "The causes for the division of labor are not always plain or autonomous, and they are located within a complex of social, economic and political relationships" (Swindell 1985:14).

Another reason why the people of Jipalom did not react to their previous experience is because historical circumstances have changed. Before, when drought struck in the 1910s and even in the 1940s, the Jola were not locked, at least not to the same degree, into a pattern of groundnut production for cash. As I indicated, groundnuts were only beginning to be experimented with during the first two decades of this century, and they were still a secondary crop in Lower Casamance by the second period of drought. As elsewhere in Africa, food security among the Jipalom population is being threatened by the introduction of a commercial crop.

Nonetheless, since the 1980s the Jipalom division of labor is moving inexorably towards a Fatiya model of gender relations. The Jola say that "rice is becoming a women's crop." By this they mean that women are progressively becoming engaged in rice cultivation without much input of labor from their husbands. Instead, husbands pay to have their wives' rice fields prepared by one of the young men's work groups. At the same time, millet is beginning to appear at the daily meals. This would have been unthinkable 30 years ago when elders used to boast that they "had never tasted millet." To this day, eating rice is greatly preferred to eating millet or sorghum.

The inhabitants of Jipalom are acutely aware that rainfall is gradually diminishing and that rice yields will never go back to what they were in the 1960s. But they recognize that shifting all their energies to growing groundnuts for cash, and buying rice in the open market with the proceeds, is not a viable alternative. Groundnut production in Senegal is stagnant, erratic, or declining (Mbodj 1992). Rice prices are no longer being subsidized by the Senegalese State. For the time being, a dual strategy of growing rice and millet for household consumption, and groundnuts for cash, seems to be the only way for families to have a minimum of food security with a measure of cash to fulfill their most urgent needs.

It is true that both systems, plateau crops and wet rice, could be made more efficient through the use of labor-saving devices like animal traction or even small, mechanized tillers (the *motoculteurs*). These devices are already being used in some areas. Moreover, projects sponsored by international agencies that are designed to build small irrigation dams to retain rainwater should be of help. But the success of these measures will depend upon the speed with which political calm returns to the now troubled region of Lower Casamance.

END NOTES

1. R. Baum (1987:1) put the estimated Jola population at about 500,000. He includes the significant numbers of Jola living in the Kombo region of the Gambia and the Cap Vert region around Dakar.

2. Recently, however, with the marked increase in the cost of imported, milled rice after the devaluation of the franc CFA in December of 1993, the men of Fatiya have been going down to the rice fields to help women till the land and carry the harvest home.

3. It is encouraging to note that in some areas, such as Esudadu, mangrove paddies were never abandoned, not even during the drought. During the 1990s, they are being cultivated even more intensively.

4. Labor invested was calculated by timing men and women with a stopwatch as they were performing a particular task in a parcel that had been, or was shortly afterwards, mapped and measured. This allowed me to calculate the surface worked – whether this was hoeing, seeding, or transplanting – and

the time it took to do the work. Although no single parcel or field ever measures more than a few hundred square meters, a household cultivates numerous parcels that together add up to a hectare or more. Hence a hectare is a convenient unit on which to perform calculations. Amounts harvested were obtained by actually weighing the rice bundles harvested from a measured field. It must be specified here that the parcels on which labor inputs were calculated were seldom the same parcels on which yields were calculated. This may or may not affect the figures, which must be taken as only approximations. The other calculations were simply derived through formulas or from existing ratios.

5. It is interesting to compare the aforementioned results with results obtained by another research team working independently along the same lines. In a study of labor time and returns to labor for lowland rice production under two regimes – transplanted rice and direct seeded rice – in the years 1983 and 1984, Posner, Kamuanga, and Lo (1991:53–54) obtained surprisingly comparable R-L values for farmers in two communities situated in comparable areas to those in which I worked. As a measure of returns to labor, they used kilograms of rice produced per day of labor invested (kg/day), rather than number of days worth of food. Because I also made these calculations (6th column in Figure 8) it is possible to compare our results. They obtained: a) 5.9 kg/day for communities near Esudadu, versus my results of 6.87 for Sambujat, and b) 8.1 kg/day for communities near the Kalunay, versus my results of 7.56 for Fatiya. However, c) they obtained 6.7 kg/day for communities near Jipalom, versus my results of 3.99 kg/day. The greatest divergence in results obtained, therefore, was in the last comparison (c). In my opinion, this discrepancy can be attributed to the fact that Posner, Kamuanga, and Lo (1991) combined yield estimates for Suel, which is near Jipalom, with figures from Tendimane, which is in a different, more productive area.

REFERENCES

Baum, Robert M.
    1987 A Religious and Social History of the Diola-Esulalu in Pre-Colonial Senegambia. Ph.D. Dissertation, Department of History, Yale University. University Microfilms International, Ann Arbor MI.
Brásio, António, ed.
    1964 Tratado Breve dos Rios de Guiné do Capo Verde. André Alvares de Almada, 1594. Lisboa: Editorial LIAM.
Cheal, David
    1989 Strategies of Resource Management in Household Economies: Moral Economy or Political Economy? In The Household Economy: Reconsidering the Domestic Mode of Production. Richard R. Wilk, ed. Pp. 11–22. Boulder CO: Westview.

Diop, Marième
  1989 Un Example de Non Insertation Urbaine: les Cas des Migrants Saisonnières de Basse Casamance à Dakar. *In* L'Insertation Urbaine des Migrants en Afrique. Philippe Antoine and S. Coulibaby, eds. Pp. 79–89. Paris: Editions de l'ORSTOM.

Glantz, Michael H.
  1987 Drought and Economic Development in Sub-Saharan Africa. *In* Drought and Hunger in Africa: Denying Famine a Future. Michael H. Glantz, ed. Pp. 37–58. Cambridge: Cambridge University.

Hyden, Goran, R. W. Kates, and B. L. Turner II
  1993 Beyond Intensification. *In* Population Growth and Agricultural Change in Africa. B. L. Turner II, G. Hyden, and R. W. Kates, eds. Pp. 401–439. Gainesville: University of Florida.

Lambert, Michael Carson
  1994 Searching Across the Divide: History, Migration, and the Experience of Place in a Multicultural Senegalese Community. Ph.D. Dissertation, Department of Anthropology, Harvard University. University Microfilms International, Ann Arbor MI.

Linares, Olga F.
  1984 Households Among the Diola of Senegal: Should Norms Enter by the Front or Back Doors? *In* Households: Comparative and Historical Studies of the Domestic Group. Robert McC. Netting, Richard R. Wilk, and Eric J. Arnould, eds. Pp. 407–445. Berkeley: University of California.
  1992 Power, Prayer and Production: The Jola of Casamance, Senegal. Cambridge: Cambridge University.
  1993 Palm Wine Versus Palm Oil: Symbolic and Economic Dimensions. *In* Tropical Forests, People and Food. M. Hladik, A. Hladik, O. F. Linares, et al., eds. Pp. 595–606. Paris: UNESCO Man in the Biosphere Series.

Louis Berger International
  1981 Program for the Development of the Baila Marigot in Casamance. Final Report. US AID.

Mbodj, Mohamed
  1992 La Crise Trentenaire de l'Economie Arachidière. *In* Sénégal: Trajectoires d'un Etat. Momar Coumba Diop, ed. Pp. 95–133. Dakar: CODESRIA, and Paris: Karthala

Mortimore, Michael J.
  1989 Adapting to Drought: Farmers, Famine and Desertification in West Africa. Cambridge: Cambridge University.

Nyerges, A. Endre
  1988 Seasonal Constraints in the Guinea Savanna: Susu Ecology in Sierra Leone. MASCA Research Papers in Science and Archaeology

[Special Issue: "Coping with Seasonal Constraints," edited by Rebecca Huss-Ashmore, with John J. Curry and Robert K. Hitchcock] 5:86–95.

Pélissier, Paul
   1966   Les Paysans du Sénégal: Les Civilizations Agraires du Cayor à la Casamance. Saint Yrieix, France: Imprimerie Fabrègue.

Posner, Joshua, M. Kamuanga, and M. Lo   ·
   1991   Lowland Cropping Systems in the Lower Casamance of Senegal: Results of Four Years of Agronomic Research (1982–1985). Michigan State University International Development Papers, Reprint No. 30.

Posner, Joshua, M. Kamuanga, and S. Sall
   1985   Les Systems des Productions en Basse Casamance. Travaux et Documents No. 4. Institut Sénégalais de Recherches Agricoles, Dakar.

Shipton, Parker
   1990   African Famines and Food Security: Anthropological Perspectives. Annual Review of Anthropology 19:353–394.

Smetsers, Maria
   1995   Gender and Agrarian Change in Senegal: Cooperation and Conflict. Paper presented at the International Congress "Agrarian Questions," Wageningen, The Netherlands, May 1995.

Swindell, Ken
   1985   Farm Labour. Cambridge: Cambridge University.

Vaughan, Megan
   1987   The Story of án African Famine: Gender and Famine in Twentieth-Century Malawi. Cambridge: Cambridge University.

Weismantel, M. J.
   1989   Making Breakfast and Raising Babies: The Zumbagua Household as Constituted Process. In The Household Economy: Reconsidering the Domestic Mode of Production. Richard R. Wilk, ed. Pp. 55–72. Boulder CO: Westview.

Wilk, Richard R., and Robert McC. Netting
   1984   Households: Changing Forms and Functions. In Households: Comparative and Historical Studies of the Domestic Group. Robert McC. Netting, Richard R. Wilk, and Eric J. Arnould, eds. Pp. 1–28. Berkeley: University of California.

Yanagisako, Sylvia J.
   1979   Family and Household: The Analysis of Domestic Groups. Annual Review of Anthropology 8:161–205.

CHAPTER 3

Indirass and the Political Ecology of Flood Recession Agriculture

Thomas K. Park
Department of Anthropology
University of Arizona

The principle thesis of this chapter is that one component of an improved agricultural development policy in Sahelian Africa should involve the elaboration of a tenure system that is based on Islamic and customary principles and rooted in an adaptation to the local environment. This is not to say that local tenure systems need no improvement, but it is to argue that the transplanting of basically European temperate climate agriculture, market economy, and individual title-based tenure is fundamentally a mistake. There are many reasons why such a system is inappropriate, but the most basic involve differences in agricultural risk, lack of literacy, and the need to have a system that can remain within the grasp of local people while adapting quickly to changing patterns of cultivation. The bulk of this chapter will focus on a discussion of what such a system might entail on the Mauritanian side of the Senegal River Basin (SRB).

A developing body of literature in anthropology calls for ethno-development, meaning development that takes into consideration and is based on continued use and development of local ethnic practice and knowledge (Scudder 1980; Spicer 1980; Stoffle, Halmo, and Stoffle 1991). The focus in this chapter will be to extend this approach into a direct critique of neoclassical economics as it has been applied to development. Neoclassical economics assumes that economic policy can be optimized only in a free market situation involving individual tenure. Despite the evidence that western economies are not based on small competitive firms and the existence of a body of theory dealing with monopoly and oligopoly situations, in the development context virtually all reference to power and institutionalized uncompetitive pricing situations gets eliminated. The alternative I argue for is based instead on the assumption that *institutionalized power is especially critical in non-literate societies* and that indigenous systems of tenure incorporate crucial local ecological and economic knowledge that, along with political democracy and free enterprise, should be incorporated into any development effort.

POLITICAL AND HISTORICAL OVERVIEW

The government of the Islamic Republic of Mauritania is dominated by Bidan (the term for members of a white North African ethnic group) whose main sources of revenue until recently were primarily in pastoralism and commerce, and secondarily in mining and off-shore fishing. The drought beginning in 1970 had a catastrophic impact on pastoralism within the Republic, and this focused the attention of the Bidan on the SRB, which had become one of the few remaining significantly productive areas in Mauritania.

Land tenure in the SRB is based on a number of competing claims. The Bidan claim much of the North bank because they conquered it at the beginning of the nineteenth century. The Haratine (a black Arabic-speaking ethnic group) claim that they as Muslims cleared land and so established rights to it (even if they did so at the urging of Bidan and paid them a portion of the crop; interpreted by Bidan as rent and alternately as protection money). The Halpulaar-en (a black Pulaar-speaking group) claim that they settled and cleared the land and cultivated it for more than a millennium (even if intermittently on the North bank). And finally, other black ethnic groups such as Soninké and Wolof claim that they too cleared and established legitimate rights to territory.

There seems little doubt both that all of these claims are based on the truth and that in practice they are often in conflict with each other. Cultivation by Haratine (who are peoples once enslaved but long since freed) is regularly interpreted in Mauritania as incapable of establishing ownership rights. Yet, in principle this is incorrect, and in practice it is recognized that this is nonsense in some areas. The principle of acquiring tenure rights through clearing unused land is well established in Islamic law (as discussed below), and Haratine are well understood to be free Muslims. One of the case studies I present (Mbout) illustrates how this is well recognized in at least one area of Mauritania.

The conflicts between Bidan and black ethnic groups such as Halpulaar-en, Soninké, or Wolof are based on fundamentally strong arguments for both sides. The French interfered in the indigenous course of events when they took over the SRB in 1891 and began to support the Halpulaar-en in the agricultural sector on the North bank because they felt the poor, ill-clothed, and low prestige Haratine were not appropriate intermediaries for the French administration. This preference opened the way for a major expansion of Halpulaar-en agriculture on the North bank in the twentieth century (compared only with the nineteenth) and led to the agricultural peripheralization of the Haratine and, secondarily, of the Bidan, who did not themselves engage in agriculture but were patrons to the Haratine. In a post-Independence Islamic perspective the French policies can be rejected as those of a *kafir* government, and the Bidan can argue that their conquests at the end of the eighteenth century still entitle them to legitimate claims in the SRB. Conversely, all those who have cultivated lands for a period of years can claim that they have legitimately established rights to the land. Even the Haratine who have been pushed into the peripheral lands by Halpulaar-en can legitimately claim that this was done illegally and with *kafir*/French assistance.

In brief, despite the current situation of conflict, we cannot incriminate any group as fundamentally wrong in their claims to land. This might suggest that the best solution would be to wipe the slate clean and begin anew. I argue instead that, although wiping the slate clean has relevance, we should not simply try to transplant a foreign tenure system into the SRB.

SYNOPSIS OF ECOLOGICAL CONSIDERATIONS

The floods of the SRB are extremely variable (Park 1992, 1993b) and cover unpredictable amounts and areas of the floodplain each year under natural conditions. Even with a controlled flood there would remain an enormous degree of variability. The key determinants of how much land is flooded for how long (one key to agricultural productivity) are the height and duration of the peak flood pulse as well as the timing and volume of the contribution to the flood from other sources such as minor tributaries and local rainfall. The result of this variability may be to flood many areas too little (the soil cannot then support a crop to maturity) or even too much (killing aerobic bacteria and lowering productivity). This is an annual variability that is quite impossible to predict, and as a result, groups of cultivators have held land in common (usually defined by some level of lineage structure) in order to have a sufficient portfolio of lands to counteract the unpredictability of the flood. More to the point in the current discussion, different lands will be cultivated in different years, and people will hold proportional rights in blocks of land from which they are allocated their appropriate share each year.

INDIRASS AND LAND TENURE

The term *indirass* embodies a legal principle that is potentially critical to development efforts if the ultimate goal is, as I believe it should be, to foster economic development within a local cultural matrix. The radical failures of development efforts in much of Africa over the last several decades have in large part been due to a general failure to facilitate growth from the bottom up. This failure represents a fundamentally misplaced confidence in elites and central government initiatives. In the Islamic Sahel, and Mauritania in particular, development efforts in the agricultural sector have consistently reflected international pressure to move agricultural production into high input irrigated systems rather than low input flood recession agriculture, which is more efficient in terms of local resources in labor and other inputs. High input agriculture has, especially in recent years, provided the basis for elite and ethnic-based appropriation of land long cultivated by others (Park, Baro, and Ngaido 1990). At a time when low input agriculture is becoming a key focus at major agricultural colleges in the United States (such as the University of Wisconsin-Madison), the continued international push for Third World countries to switch to high input agriculture seems anachronistic and is either excessively naive or exhibits a fine disregard for the interests of Third World countries. In Mauritania, the population is small and the sources of foreign exchange are few. High input agriculture, whose only advantage is higher output per hectare, seems particularly inappropriate in this context given the major costs at which this increased output can be obtained in the best of years, the enormous debts incurred in poor years, and the unacceptable level of reliance on foreign exchange for purchases of inputs of fertilizer, machinery, and fuel for the irrigated perimeter crops that such agriculture entails.

In this chapter, I take as my starting point for reconceptualizing the relationship between flood recession ecology and the larger socio-cultural system in the SRB the Islamic term *indirass*. *Indirass* is an Arabic word, derived from the root d-r-s whose base meaning today is "to study." In classical texts, however, the base meaning was linked to "effacement" or "erasure." Lane (1984 [1863]:870) provides the example *indarasa rasm* meaning a trace (*rasm*) that "became covered with sand and dust blown over it by the wind." More generally, the verb itself contained the implication of nature erasing or covering up human-made traces. The modern association with studying, presumably, would have been via the student's slate, which had to be continually wiped clean to make room for further exercises. In legal terms, *indirass* is now used to refer to the obliteration of traces of cultivation (or habitation) due to the passage of time. The significance of this term is closely tied to basic principles of Islamic law, the *shari'a*.

In Islamic law oral testimony has priority over written testimony (Wakin 1972). This preference reflects a fundamental prioritizing of the current Islamic community over all past Islamic communities. Unlike in Europe, where old documents might be used to kick someone off land they have long occupied, in the Islamic world, under *shari'a* law, old documents are worthless unless connected by a series of documents demonstrating continued ownership of the property in question and capped by current oral testimony to the effect that the person still is recognized as the owner (Amar 1913:78–79). In reality only the last, the oral testimony, has any real significance and the earlier documents have value only if two competing claims are both supported by current oral testimony by impeachable witnesses. In the general situation, current witnesses to ownership take precedence over all past claims. The law thus supports the claims of the current community over those of any past owners. A document of sale is simply a written embodiment of an oral attestation that the seller has been in possession of the property for the last ten years or the last one year plus the attestation that any former owner is unknown (in the Malikite system). As such it stands or falls on the oral testimony documented in the bill of sale. A challenge focuses exclusively on the testimony, and ownership itself does not require any written documentation.

The effect of this basic principle is to de-reify ownership rights. Ownership becomes an explicitly social phenomenon, which has no independent significance and changes along with society. To some extent this is true in the European system, but the linkage with current society in the Islamic system is far more direct and immediate. In the Islamic system, a brief period of time, usually from three to ten years, without the exercise of ownership rights is sufficient for those rights to revert, on the principle of *indirass*, to the Islamic community. Perhaps even more importantly, rights are maintained through economic practice, as confirmed by neighbors and other members of the community, rather than through careful written documentation.

This last point has enormous significance in situations where simple private property does not prevail. To the extent that rights are themselves tied to practice, subsidiary rights if regularly exercised become as real as any rights. Thus the right in the SRB of someone who has traditionally cultivated land in return for paying a small tithe becomes as real as that of someone who has land on which they pay nothing. There may be a prioritizing scheme that puts the latter before the former when land is short, but the former still has rights that take precedence over people who have not cultivated the land in the most recent period.

This system has particular importance for women and the poor. In the SRB, women often have rights to use the land but may have to pass those rights on to their sons (and hence not be considered full owners). In the case of the poor, individuals may have traditional rights to cultivate land but are obligated to pay various tithes and fees to maintain them. When such complex tenure situations are mapped onto a simple ownership list, as in the context of a development project, they normally disappear, whereas when based on practice, they are easily maintained as confirmed by oral testimony.

IMPLICATIONS FOR DEVELOPMENT

The flexibility of the traditional system and its tight linkage to current practice are, in the context of a society in which literacy is far from universal and written record keeping is far less than perfect, a vast improvement over the sorts of simplistic and unrealistic systems of ownership regularly proposed for development projects. The latter tend to organize ownership around registered parcels, which is a system that has several fundamental flaws. First, the lists always fail to keep pace with social change. Second, the lists ignore the complex practice of agriculture and simplify tenure rights to the general detriment of women and lesser right holders. And, third, due to the implicit literacy requirements of the system, the lists take ownership out of the hands of the community and put it in the control of elites. Traditional, oral testimony and practice-based systems seem, in contrast, models of justice.

In 1983 and 1984, the Islamic Republic of Mauritania enacted two pieces of legislation on land tenure. These were the Ordinance 83.129 (5 June 1983) and Decree No. 84.009 (19 January 1984). The latter was a more detailed implementation document for the former. Article 9 of the Ordinance states:

- Dead lands (*terres mortes*) are the property of the state. Lands which have never been developed or whose development has left no trace are considered dead.
- Extinction of property rights by *indirass* can be opposed by the original proprietor and by his heirs, but not in the case of properties which have (since) been (officially) registered (by someone else).

It should be noted that the category of "dead lands" is itself a traditional category in Islamic law. To simplify egregiously (ignoring *shi'ite* differences in particular), Islamic property law has traditionally classified land in one of two ways; according to the way the land was acquired by the Muslim community (i.e., conquest, cession without violent resistance, reclamation, etc.) or according to the type of taxes paid on the land (Tabataba'i 1983). The former was most significant in the central Islamic lands where large non-Muslim minorities were incorporated into the early Islamic states, while the classification according to type of tax was more typical in North Africa and the Sahel where almost universal conversion to Islam occurred over significantly briefer periods. In these areas the concept of dead lands seems to have simply included all lands not in use. Taxation or the extraction of some form of tithes was linked throughout the Islamic world to the payment of religious duties, specifically *zakat*, the payment of alms to the poor, and *'ushr*, a tax of one-tenth on agricultural production for all but the abjectly poor.

In the central Islamic lands *'ushr* was paid by those not paying *kharaj*, a land tax required on lands that had originally (when the local classification was drawn up) belonged to non-Muslims (Tabataba'i 1983). The latter was put at a higher rate than the 10%

normal for 'ushr. In North Africa, the payment of taxes itself tended to define the land. Even in Morocco, where the center's claim to tax was not always accepted, payment of 'ushr was generally accepted by the population as the legal due of a justly run state. Since payments were a portion of production, unproductive land did not pay anything and thus, in practice, after a period of years, fell into the category of dead lands without recognized ownership. The question of who owned dead lands gave rise to some controversy in Islamic law, but in North Africa and the western Sahel where Maliki law was the norm, the Muwatta of Imam Malik ibn Anas was accepted as the key text, and this, in Section 36.24 paragraphs 26 and 27, states quite clearly that "If anyone revives dead land it belongs to him" (Malik ibn Anas 1989:307). The alternative favored in some Islamic lands was to have it belong to the state and thus have to be purchased or leased from the state.

In the SRB, the classification of land as dead has one significant problem. In the context of an irregular flood and flood recession agriculture much of the prime agricultural land is, in the normal course of things, not used for many consecutive years. The concept of dead lands as defined in agricultural systems elsewhere under Islamic law (such as those lands unused for ten years) would, if not modified, wreak havoc in the traditional land tenure system. At the same time, the prioritizing of current social practice and oral testimony is not a problem in the flood recession context. Nor is the concept of dead land itself since most people base their claims to land on having, as free Muslims, cleared the land from brush (with the implication that it was abandoned or never used and hence "dead land"). The two case studies discussed below illustrate how local systems have approached the concept of indirass.

CASE STUDIES

Mbout

Mbout (see Figure 1) is an administrative center located on the Gorgol Noir (a tributary of the Senegal River) just above the Foum Gleita Dam. The Foum Gleita reservoir now floods most of the land formerly cultivated in the region while its downstream irrigated perimeters have substituted for lands now under the lake or put out of production below the dam. Before construction of the dam and gravity fed perimeter, approximately 60% of the land was owned by Haratine.

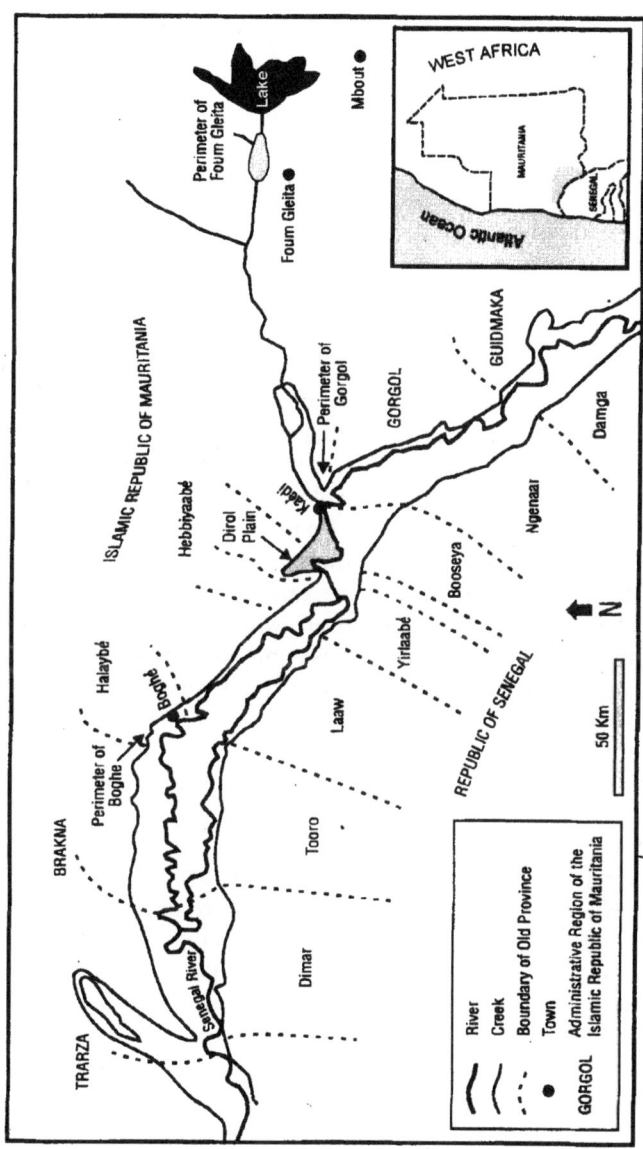

Figure 1. The middle Senegal Valley, showing the locations of the case study sites.

The local Bidan *qadi* in 1986 provided ample documentation that the Haratine had over the years sold parcels of land at market prices to local Bidan and justified this evidence of the Haratine ownership of land in terms of Islamic law; they had come to the region, admittedly at Bidan urging, in the late eighteenth and early nineteenth centuries and on arrival had cleared long fallow land. As free Muslims this gave them clear title to the land and subsequent ability to sell it. These rights were obtained whether or not the Haratine paid some tithes to local Bidan elites. In the *qadi*'s view, such payments were considered as protection monies paid for protection from slave raiders and were in no way to be interpreted as rents (which would carry the implication that the Haratine were tenants rather than owners), though he acknowledged that this was a common interpretation in other parts of Mauritania.

In this region cultivation was traditionally based on flood recession agriculture, though the lands cultivated comprised more narrowly de-limited areas than in the Senegal River valley, and consequently, there was minimal emphasis on annual reallocation. More important, there were only two ethnic groups having legal rights in the arable land, the Haratine and Bidan, and the latter did not themselves cultivate. Bidan arranged to have cultivation done by others, either Haratine or Peul (a pastoral group of the Halpulaar-en, who no longer had enough livestock to remain pastoralists, as was common in the years following the Sahelian drought of the 1970s). The result was that Haratine did the great bulk of the cultivation. The majority of Haratine held contiguous blocks of land in common at the fraction level (a named group of relatives and affines), yet within some parts of the Mbout region individual tenure prevailed, and within the contiguous blocks of land there was far more continuity in the cultivation of specific plots by specific individuals than in the Senegal River valley itself.

The narrowness of the flood plain, its relative uniformity, the low population density, and the non-stratified character of Haratine society contributed to a fairly unique tenure situation (for Mauritania). In this region, the Islamic principle of *indirass* was not only fully accepted, in the sense that abandonment of a piece of land in favor of another for a period of time was interpreted as the relinquishment of all rights to the land, but the period considered sufficient to establish the extinction of former rights was a mere three years. The local *qadi* suggested that this period was justified by the ecology of the Sahel and the characteristics of local agriculture; it took only three years to extinguish all traces of cultivation.

It is clear that in the years before the development of the Foum Gleita reservoir, the principle of *indirass*, applied after three years of non-use, benefitted most prominently the local Haratine community because it was they who engaged in all cultivation.[1] It is likely as well that one reason for a three-year period may have been a combination of relatively low population levels, political exigencies, and the relative unimportance of social stratification. In the twentieth century, according to administrative records and the *qadi* of Mbout, who had been in office in 1986 for over 40 years, a number of groups were settled in the area as rewards for service to the colonial state. The most prominent were the former Tirailleurs Senegalais. This group is still distinguished by name, though in practice they are distinct from other Haratine primarily in the location of their lands and the maintenance of different oral traditions. A brief period to qualify land as "dead" would obviously have facilitated the distribution of lands to new immigrants. The low population density would have made this distribution feasible. The colonial archives in Mbout are full of records indicating that the colonial administration felt that the area was seriously underpopulated right through the 1950s and, in consequence, adopted a policy of encouraging settlement in the area by cultivators and increased use of the area by nomads. The lack of social hierarchy within agriculture, in the immediate Mbout region, would itself have been conducive to the establishment of new groups on an equal basis.

In short, this particular pre-irrigated development pattern clearly reflects local ecology, demography, and political processes. It contrasts in a number of ways with the following case, which is perhaps more typical of the more highly populated areas of the Senegal River valley, but nevertheless exhibits a number of commonalities as well. Most importantly, oral testimony provides the crucial determinant of tenure, and the current practice of the members of the Islamic community, in all its complexity, is the fundamental datum in the administration of justice and the structure of the economy. Prioritizing the rights obtained by clearing and continued cultivating of land over and above claims due to past conquest (by Bidan groups) or other claims, such as religious or military prominence in the region as evidenced by payments received, clearly took both great wisdom and erudition by the *qadi*.

Boghé

In the Boghé region (Figure 1), current patterns of tenure date from the mid-nineteenth century because at the end of the eighteenth century the Emirate of Brakna took over this section of the right (North) bank of the Senegal River, and the Halpulaar-en and Peul fled to the left (South) bank. Subsequently, many of them made arrangements to come back to cultivate some of the North bank lands, but numerous Haratine were now established in these lands by the Emirate of Brakna (similar arrangements were made by the Emirate of Trarza to the west). After the French conquered the valley in 1891, they ruled, for complicated reasons (Park 1993a), that all holdings in the Boghé area on the North bank cultivated by Halpulaar-en were held only in usufruct and could not be claimed as legal property. In principle this ruling held until Independence in 1960. In practice, Halpulaar-en cultivators continued to elaborate agricultural holdings on traditional, pre-Emirate conquest principles, combined with Islamic law, throughout the colonial period. Yet what this meant after Independence was that at least three groups could legitimately, although using different criteria, claim rights to land in this region: Bidan members of the former Brakna Emirate (through the right of conquest), Haratine (by clearing unused land and cultivating it as free Muslims – even if at the instigation of Bidan and even if they continued to pay tithes), and Halpulaar-en (by clearing land and cultivating it for decades as free Muslims without paying any tithes to the Emirate of Brakna after the French conquest, because the French took over jurisdiction and simply insisted that land was owned in usufruct only. In retrospect, their rulings, as those of *kuffār*, could be ignored after the colonial period).

Within the flood recession context, the broad flood plain was divided into pieces of common property called *colladé* and held by lineage groups (Halpulaar-en) or fractions (Haratine). These lands provided a large portfolio from which annual allocations to members were made on the basis of the year's flood pattern and the shares held by individual members. This practice has a significant advantage over individual tenure, which would regularly result in individuals finding themselves with land that was particularly poor or even unusable. Because the good years can support far more than the long-term average population, property holding groups own more land than they can cultivate in good years and so have elaborated ways of subleasing lands to peripheral members of the community, such as pastoralists, Haratine, the poor, and lower castes. These arrangements have become institutionalized and amount to the recognition of subsidiary tenure rights.

Traditionally, disputes can arise over rights to land between individuals with subsidiary tenure rights and those with more direct rights. The *qadi*'s court in Boghé preserves records of cases between many such disputants. In one case a woman who had for years cultivated some land that had been unused suddenly had the land taken away from her by a man who claimed ownership. This dispute was initially taken to the *qadi* for presentation but was finally arbitrated at the village level (at a village a few kilometers west of Boghé) using customary law. The case is typical, in that, although the legal principle of *indirass* would suggest that someone who cultivated unchallenged unused land for a period of years would, in so doing, acquire rights to the land, the ecology of flood recession agriculture forces communities to leave some lands fallow, or used only by peripheral members (due to their low productivity) for many years simply because the flood has rendered other lands much more productive. The flood pattern, however, can quickly change and reverse the order of productivity for a group's lands. In this case the village in question was one in which a large number of the inhabitants were fairly recently settled Peul (FulBe – traditional nomads who belong to the same basic ethnic group as the sedentary Toucouleur and together comprise the Halpulaar-en). As a result, the claims for long residence and cultivation of the land were particularly doubtful, and this made the resolution of the matter at the village level all the more critical. A negative judgement at the *qadi*'s court against the man who claimed past ownership could have inconvenienced a lot of people. The principles of the *shari'a*, though the basis for most ethnic claims to land (the group members cleared the lands and so established their rights), do not lend themselves directly to the arguments favored by the local elites and so these have traditionally forced parties to settle disputes over land through traditional arbitration.

THE RECENT EVENTS

The recent events in Mauritania, involving the dislocation of 400,000 people and the expulsion of the greater part of this number of Halpulaar-en from the SRB stemming from an incident in April 1989, substantiate the pessimistic view that loopholes in the law will be taken advantage of by elites if the incentive is adequate. In this case the drought had severely reduced the viability of pastoral modes of production in Mauritania. Simultaneously, major development efforts including construction of the Diama Dam, to prevent the entry of a salt-water tongue beyond a certain point in the Senegal River, and the Manantali Dam in Mali, to provide hydroelectric power and to make possible controlled irrigation along the Middle Senegal River, made the area of the SRB look increasingly attractive. The net result (Park, Baro, and Ngaido 1990) was that Bidan elites began to push for modifications in the land tenure legislation to facilitate access to lands in the Middle SRB for private development schemes.

The first legislative efforts involved Circulaire, which were nominally intended to clarify parts of the legislation in force (Park, Baro, and Ngaido 1990). In fact these efforts did far more. As a group they set the ground for legal ways of short-circuiting the protections built into the Mauritanian judicial code. In particular, the legislation of 1983–84 provided a series of steps that had to be followed before land deemed vacant (dead) could be acquired by the state or a private party. Three steps in this process were particularly important. The first was acquisition of vacant ownerless lands by the state according to *shari'a* (*indirass* declared after the appropriate delay: Article 11 of Ordinance No. 83.127). The second was public notification of the impending change in status to a one-month concession to a new occupant (Article 29 of Decree 84.009), followed by a further temporary concession if a series of additional steps were taken. The third and final step was a definitive concession if the agreed upon investments were made. The changes instituted by the Circulaires included increasing the efficiency of the process by moving all land concessions of a "non-major" character (undefined) from the jurisdiction of high-level bureaucrats to that of local bureaucrats such as Préfets. They further justified the practice of making concessions from land that was in no real sense vacant, circumventing the provisions of the *shari'a* concerning dead lands, ignoring the need for witnesses to the effect that the land was indeed without owner for a given period of years, and condoning the practice of not even bothering to inform the public of the impending concession (Park, Baro, and Ngaido 1990).

The result was expropriation of land on a massive scale. Cultivators were ejected from their fields in mid-harvest on the grounds that the lands now belonged to Bidan groups, even though the principle of *indirass* could not even begin to be said to apply and so the land could not become domain or state land available for concessions. Innumerable villages of Halpulaar-en were expelled from their country, and some elites simply took whatever lands struck their fancy and imprisoned, killed, or expelled all who objected. Amnesty International, Africa Watch, and other human rights organizations have charged the Mauritanian government with explicit participation in and support for grave crimes against humanity (Park, Baro, and Ngaido 1990).

HISTORICAL CONTEXTUALIZATION OF ISLAMIC LAW

Exploitation of the weak by the powerful is a well-entrenched historical practice in virtually all parts of the globe. The Islamic world is no exception. There, as in other countries, the legal system has been systematically warped to fit the needs of the powerful. Johansen (1988) has made the case quite convincingly that Islamic law and land tax were systematically reinterpreted in the Mamluk and Ottoman periods to benefit the wealthy and deprive the poor of all legal property rights. The Mauritanian case is, thus, not qualitatively different either from the Ottoman case or from the tenure changes exemplified by the enclosures in Great Britain when the rights the poor had held under the feudal period were systematically expropriated to benefit the rising industrial and landlord classes.

The corruption of elites should not, however, distract us from the potential value of an oral testimony-based legal system such as the *shari'a* in the context of development efforts in societies where literacy and careful written record keeping are not likely to be the norm for many decades. The flexibility of such a system, especially in the context of an annually changing usable flood plain, can scarcely be exaggerated.

DISCUSSION

The two case studies presented demonstrate that political and ecological considerations enter into the application of Islamic principles. Rather than suggesting that one set of rigid principles can profitably be substituted for another, the argument I put forward relies on the flexibility of the application of principles based on oral testimony and current community practice when compared with attempts in the late twentieth century to apply principles which, although they may work in thoroughly literate societies, have no special claim to justice or efficiency and are completely unfeasible in the current impoverished African context.

Land registration schemes in Africa have shown little long-term benefit, and the registration lists have provided increasingly inaccurate pictures of reality as the schemes age and fail to reflect changes due to demography and market conditions. Impoverished conditions, generally low levels of education, and numerous informal economic arrangements do not meld well with registration schemes designed to prevent land fragmentation, and which are based on simplistic conceptions of traditional tenure rights.

The recent rhetoric about free-enterprise and individual tenure substitutes dogma for the complexity of reality. There are indeed many advantages to competition and decentralized decision-making, the real virtues of a capitalist economy. Western economies, however, have not relied exclusively on competition, and since the Depression western agriculture has used subsidies, bailouts, and tax breaks to counteract the unpredictability of agricultural production. It is thus nothing but rhetoric when the International Monetary Fund (IMF) suggests that all problems can be solved by promoting individual competitive farms in agriculture. Although in some areas the adoption of such a policy would change little, at least in the short run, in other areas such as the SRB under a flood recession regime it would result in catastrophic change in the short and long terms.

An issue that has contributed significantly to neoclassical dismissals of traditional tenure systems is the claim that they provide disincentives to investment in comparison to an individual tenure market-oriented system. There is certainly some truth to this insofar as annually reallocated common property discourages individual investment in the property, though it provides no disincentive to collective investments. In addition, people working land to which they have less than full ownership in fee simple rights (which includes most individuals, particularly the poor and women) will have few incentives to invest in capital development. Nevertheless, a similar criticism of the capitalist system can be made. To the extent that workers are not owners, they have less incentive to maintain the capital equipment (whether it is land or machinery). In the US more than 90% of equity is owned by a mere 10% of the population, and this must unquestionably contribute to the degradation of land and equipment, and the less than full use of the talents of the laboring population. The further failure of the system to charge full social costs (i.e., for pollution) to industry has well known impacts on the environment that will only be aggravated by any increased competition from the lowering of international trade barriers. In short, the incentives for investment (comprising maintenance and new investment) are less than ideal in both systems. The solution to both problems is undoubtedly to be found in the legislative area; whether it involves securing compensation for individuals making investments (something basically already supported in Islamic law but which could do with some improvements) or international legislation to adjust tariffs to compensate for pollution control failures.

The typical development approach suggests that agricultural policy should be shaped by estimates of short-term comparative advantage, and the economy should be structured solely in terms of individual tenure and competition. If we take Mauritania as an example, rice cultivation in irrigated perimeters has been pushed because it is a product with a world market value that Mauritania can produce. Yet, it is clear that Mauritania does not in any real sense have a comparative advantage in the production of rice – if the national cost of the foreign exchange needed to buy fossil fuels for the pumps, fertilizers for the rice, and machinery of all types is included. No study has provided a remotely persuasive argument to show how the foreign exchange earnings will fully cover the foreign exchange costs, let alone bring in significant profits. It is true, however, that a private party subsidized by the state in terms of the provision of free land, cheap credit, and administrative help could turn a profit producing rice in good years. If all implicit subsidies are excluded and the typical range of climatic conditions is included, the figures would appear quite different. All is relative apparently in IMF eyes, and countries are to do their best to produce something of benefit to world trade – even if alternatives are available that are preferable from a national perspective.

By contrast, indigenous sorghum, though it does not have high value on the world market, can be grown with flood recession agriculture at no foreign exchange cost to the country and is undoubtedly much less risky both politically and economically. Controlled release of floods by the Manantali Dam could significantly decrease the risk and improve production, yet the obvious advantages count for little because they do not entail adoption of individual tenure, a free enterprise system, and capitalist production for an international market.

The alternative here proposed would take advantage of the provisions in Islamic law for oral testimony supporting a wide spectrum of rights held by men and women, and similar provisions for the prioritizing of the rights of the current Muslim community, via appropriate application of the principle of *indirass* to formulate development policy that neither insists on individual tenure in all places and times nor focuses on short-term comparative advantage. The difficulty with most criticism of current development policy is that it does not go far enough. The virtues of free enterprise get systematically embedded in a complex of extraneous dogma that is never properly analyzed or unpacked. Democracy and free enterprise do not imply that there is no role for anything but individual tenure or that the state should adopt a complete laissez-faire attitude. It is unlikely that any state has ever fully adopted a laissez-faire policy and certainly no state today even begins to approach such a position. Most areas of the economy in modern states are run by corporations that have multi-national status, at least oligopolistic market positions, and major tax breaks. Only a small sector of the economy remains even remotely composed of small competitive firms. It is thus extraordinarily hypocritical for western organizations like the IMF to expect Third World countries to adopt an economic policy that assumes their economy and other economies are comprised exclusively of small competitive firms competing at no disadvantage with all the world's firms. Instead, the reality of market power should be accepted, and the development policies of Third World countries, while encouraging democracy and free enterprise, should focus first on national interest and only secondarily, and with critical awareness of the inequities of the world system, on production for a world market. This might in some cases involve encouraging production systems that do not assume a continuous and appropriate supply of credit, logistic supplies, and labor, but do assume that the imperfections in the world economic system dictate a measure of self reliance and the adoption of legal and economic policies appropriate to local communities and local ecologies.

END NOTE

1. I have discussed in detail the political ecology of the region in the context of the irrigated perimeter at Foum Gleita in Park (1988).

REFERENCES

Amar, Emile
    1913   L'Organisation de la Propriété Foncière au Maroc. Paris: Paul
           Geuthner.
Johansen, Baber
    1988   The Islamic Law on Land Tax and Rent. London: Croom Helm.
Lane, E.W.
    1984   [1863] Arabic-English Lexicon. Vol. 1. Cambridge: The Islamic
           Texts Society.
Malik ibn Anas, Imam
    1989   Al-Muwatta of Imam Malik ibn Anas. Translated by Aisha Abdur-
           rahman Bewley. London: Kegan Paul.

Park, Thomas K.
  1988 Indigenous Responses to Economic Development in Mauritania.
        Urban Anthropology and Studies of Cultural Systems and World
        Economic Development 17(1): 53–74.
  1992 Early Trends Toward Class Stratification: Chaos, Common Prop-
        erty, and Flood Recession Agriculture. American Anthropologist
        94(1): 90–117.
  1993a Arid Lands and the Political Economy of Flood Recession Agricul-
        ture in Fuuta Tooro. *In* Risk and Tenure in Arid Lands: The Politi-
        cal Ecology of Development in the Senegal River Basin. Thomas K.
        Park, ed. Pp. 1–30. Tucson: University of Arizona.

Park, Thomas K., ed.
  1993b Risk and Tenure in Arid Lands. The Political Ecology of Develop-
        ment in the Senegal River Basin. Tucson: University of Arizona.

Park, Thomas K., Mamadou Baro, and Tidiane Ngaido
  1990 Conflicts Over Land and the Crisis of Nationalism in Mauritania.
        Report for the Land Tenure Center, University of Wisconsin-
        Madison and USAID. LTC Paper No. 142.

Scudder, Thayer
  1980 River-Basin Development and Local Initiative in African Savanna
        Environments. *In* Human Ecology in Savanna Environments. David
        R. Harris, ed. Pp. 383–405. London: Academic.

Spicer, Edward H.
  1980 The Yaquis: A Cultural History. Tucson: University of Arizona.

Stoffle, Richard W., David B. Halmo, and Brent W. Stoffle
  1991 Inappropriate Management of an Appropriate Technology: A Re-
        study of Mithrax Crab Mariculture in the Dominican Republic. *In*
        Small-Scale Fishery Development: Sociocultural Perspectives. John
        J. Poggie and Richard B. Pollnac, eds. Pp. 131–157. Providence:
        University of Rhode Island, International Center for Marine Re-
        source Development.

Tabataba'i, Hossein Modarressi
  1983 Kharaj in Islamic Law. London: Anchor.

Wakin, Jeanette
  1972 The Function of Documents in Islamic Law. Albany: State Univer-
        sity of New York.

CHAPTER 4

The Ecology of Food Security in the Northern Senegal Wetlands

John Magistro
Environmental and Societal Impacts Group
National Center For Atmospheric Research

Indigenous systems of knowledge and practice, in particular peasant livelihoods in farming, herding, and fishing, have gained attention among social scientists in recent years for their capacity to sustain agrarian communities, even when altered by forces of political and economic change (Brokensha et al. 1980; Chambers 1983; Richards 1985; Altieri 1987; Altieri and Hecht 1990). That local environments can be sustainably managed by resident populations has been demonstrated in contexts as diverse as indigenous pastoralism, forestry, and farming (Little 1984; Cernea 1985; Dyson-Hudson 1985; Horowitz 1985; Little and Brokensha 1985). In riparian settings in Africa, Scudder (1980, 1988, 1989, 1991), Adams (1983, 1986, 1987), and Horowitz et al. (1991) have persuasively documented the ecological viability of existing river basin production regimes. Although none of these studies portrays local farming systems as the flawless panacea to food-shortage ills in developing nations, they nevertheless recognize them as viable production assemblages to be built upon rather than abandoned. In Africa, riverine ecosystems particularly merit further study as they are increasingly the target of multinational development initiatives to generate hydroelectric power and facilitate irrigated agriculture. Such development projects alter stream flow and the seasonal discharge of floodwaters, inevitably threatening the food security of downstream populations. In consequence, riverine communities that have derived subsistence for centuries from a diverse range of agropastoral and fishing activities are suddenly being forced to abandon old ways in favor of more modern, capital-intensive strategies of production. In making the transition from customary to externally imposed livelihoods, smallholders are compelled to choose between strategies of diversification and homogenization – that is, between traditional low-input, variegated systems of multiple resource management and poly-cropping, on the one hand, and high-input, capital-and-labor intensive regimens of commercial monoculture, on the other.

In this chapter, I discuss an example from the Senegal River Valley (Figure 1) of riparian dwellers in a seasonal wetland habitat who face the potential loss of a centuries-old practice of recession farming, floodplain fishing, and dry-season bottomland herding. At the turn of the century, this region was a net exporter of millet and other cereals to neighboring provinces (Minvielle 1985:55). Now, the inhabitants of the valley are experiencing food shortfalls, economic dislocation, environmental decline, and vulnerability to drought and famine.

This situation can only worsen. By the year 2015, the region's total food needs for grain may predictably double (GERSAR/CACG et al. 1990). This rising demand, which characterizes the entire post-independence era in Senegal, has been met by protracted periods of drought in the West African Sahel and the progressive failure of the agricultural sector to meet the new demand. Energy price hikes of the early 1970s and a steady decline in the world market price of groundnuts, which are a major source of foreign revenue for the Senegalese economy, caused the national deficit to balloon, absorbing 15.8% of the gross domestic product by 1980 (Duruflé 1988:28). In parallel with economic decline, public demand for imports rose, as evidenced by the consumption of Asian-produced rice as the nation's dietary staple. By the mid-1980s, Senegal was producing only 52% of its total food requirements, the remainder coming from purchased imports at 40% and food aid at 8% (Commander et al. 1989:154). Agricultural output has been unable to keep pace with an annual population growth rate

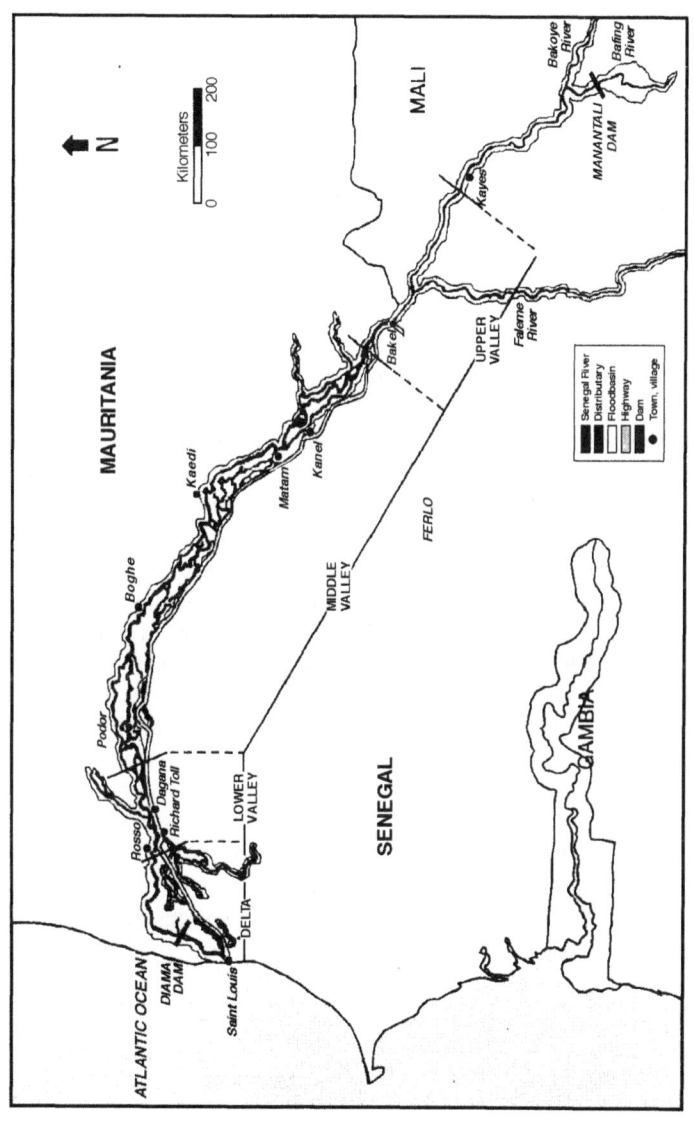

Figure 1. The Senegal River Valley.

approaching 3%. Consequently, the Senegalese government has come to rely heavily on the support of foreign donors, who accounted for 87% of the government's investment budget by 1986–1987 (ibid.:150).

Spurred by economic decline, the Senegalese government embarked on an ambitious program of development in the Senegal River Valley. Intent on stemming the growing domestic demand for imported rice, the government instituted policy measures to transform the valley floodplains into a regional grain basket capable of provisioning the hinterland and simultaneously providing enough surplus to meet the food demands of the country's growing urban population. This task was to be done in conjunction with the neighboring states of Mali and Mauritania, through the creation in 1972 of the Senegal River Valley Development Authority (the OMVS). This organization, supported by multilateral donor assistance, was to transform the Senegal River Basin (SRB), via dams and pump-irrigation schemes, into a major rice producing region.

The initial step in fulfilling this goal was realized in 1986–1987 with the completion of two construction projects, the Diama Dam situated in the lower delta of the valley and the Manantali Dam located in southwestern Mali on one of the Senegal River tributaries (Figure 1). These dams were built, first, to provide hydroelectricity to cities such as Dakar, the capital of Senegal, where energy demands are outstripping supply. Second, they were to facilitate the development of irrigated rice throughout the SRB (240,000 ha in Senegal, 126,000 ha in Mauritania, and 9,000 ha in Mali) by stabilizing seasonal fluctuations in the river level in order to make double cropping possible. Third, they were to enable the creation of a navigable watercourse for riverine commerce including development of an internal port of trade at Kaye, Mali.

For more than two decades, the governments of Senegal and Mauritania have been developing rice irrigation perimeters on both sides of the river. As of 1988, development of an estimated 39,270 ha had been completed in Senegal and another 16,856 ha in Mauritania (Woodhouse and Ndiaye 1990:4). Other intermediate goals include expansion of rice production through double cropping and the installation of water turbines for electricity generation at Manantali. In the meantime, a transitional phase involving the conversion to full-scale rice irrigation is being planned. This phase would wean producers from customary forms of farming, fishing, and herding on the floodplain. By the year 2000, resident smallholders, forced to abandon a time-tested form of recession cultivation, could become full-time irrigated rice farmers.

Similarly, many of the world's wetlands in both developing and industrial nations are at risk, as economic growth, agricultural development, and human encroachment take their toll. Until recently, however, few efforts have been made to examine the social and political costs of undermining the ecological stability of tropical wetland regimes (Scudder 1988, 1989; Horowitz 1989; Horowitz et al. 1991). In this chapter, I will build upon the work of those cited here, focusing on the social and environmental outcomes of management change in a comparative sample of three wetland communities affected by the impoundment of the Senegal River. I begin with the idea that anthropological research over the years has demonstrated that diversified, customary farming systems are often more viable and less prone to environmental risk than are homogenized, capital-intensive monocultures (Richards 1985). In examining the case material presented here, I suggest that *riparian communities in the middle Senegal Valley will be better able to provision households with adequate quantities of food through preservation rather than elimination of an array of cropping, herding, and fishing activities.* My comparative data on labor time allocation and crop yields further demonstrate that *the loss of customary activities in favor of irrigated rice farming lowers food security and undermines the social and economic reproduction of flood-dependent agropastoral and fishing communities.*

The ongoing quandary of regional food security is illustrated in this chapter by examining the farming practices of smallholders and quantifying labor input versus grain output for each wetland system. Data were gathered in a comparative study of three wetland communities all located within 10 km of one another in the Senegal Valley district of Matam (Figure 2). Data from the three sites particularly illustrate the local variability in access to environmental and human resources among small communities in the valley. Thus, the primary site of Thiemping – located approximately 17 km upstream from Matam, the administrative seat in the region – was chosen for its population size, sociocultural and occupational heterogeneity, and broad mix of agropastoral and fishing activities. Thiemping exemplifies the extreme high degree of human and resource diversity in Senegal River Valley communities. Here, smallholders exploit an array of cropping systems that includes recession cultivation in three distinct habitats, two systems of irrigated horticulture, and a rainfed farm regime.

This rich assemblage of productive activities contrasts with the modest land holdings at two nearby secondary sites, Thially and Dembaka, which were chosen to complement the primary site and provide

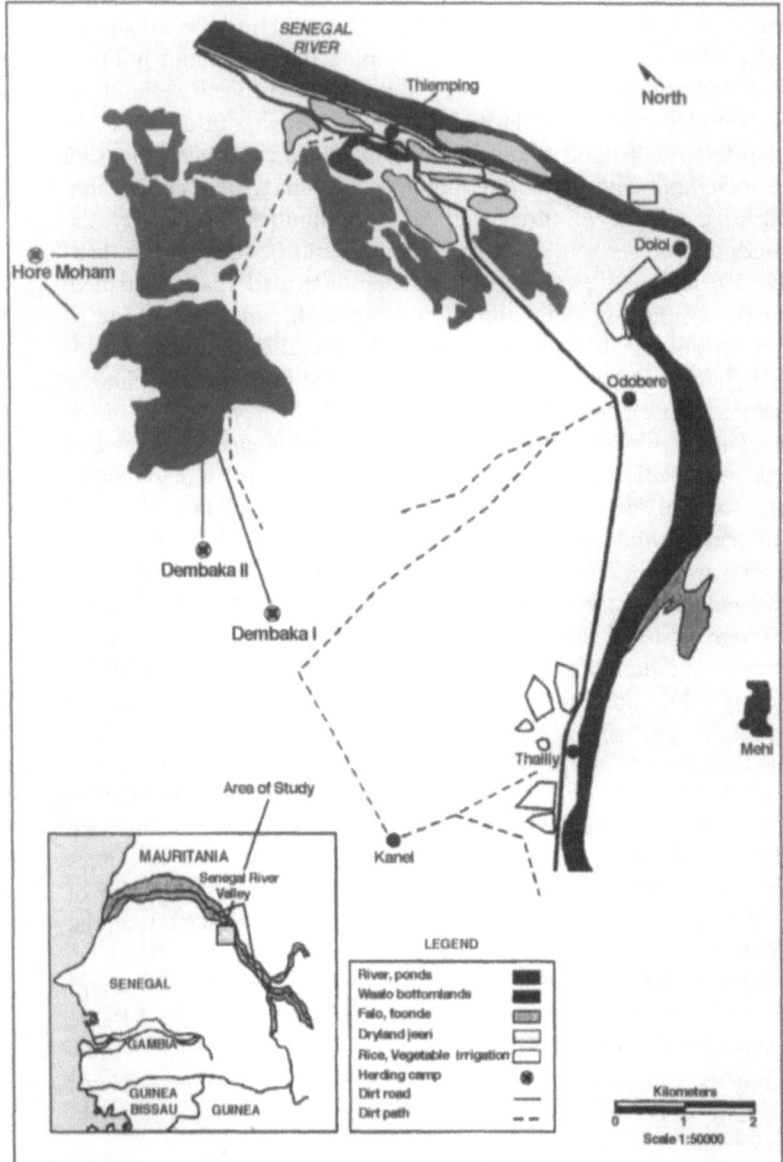

Figure 2. Area of research in the middle Senegal Valley district of Matam, showing the locations of the Thially, Thiemping, and Dembaka study villages.

a basis for comparisons of social structure, demography, and productive activities. The secondary sites have socioeconomic and resource configurations that are more uniform than those at Thiemping. For example, at Thially, smallholders farm under conditions of resource constraint, relying almost exclusively on irrigated rice agriculture.

The smallest settlement in the study, Dembaka, characterizes the many seasonal encampments of transhumant FulBe pastoralists that dot the southern fringe of the floodplain. Similar to Thially smallholders, the production profile of the Dembaka herder is homogeneous. These agropastoralists rely on dry season access to bottomland pastures and the cultivation of recession lands as sharecroppers. Dembaka herders, however, interact symbiotically with sedentary agriculturalists such as those at Thiemping. At all sites, data gathered during the 1988–1990 research period include household farm production and commercial activities, patterns of income generation and spending, and food consumption regimens.

In what follows, I briefly describe the regional history and social organization of the populations that inhabit the middle Senegal Valley and sketch the essential environmental features of an ephemeral Sahelian wetland habitat. These factors underlie the central theme of the chapter – an inter-village assessment and comparison of the ecology of food security at the three research sites. The chapter concludes with a discussion of the implications of the research findings for policy, including a call for a more anthropologically apprised approach in the formulation of river basin development policy in Africa – one that is more receptive to the voices of the unheard, or those who have been aptly described as the "victims of development" (Horowitz 1989).

THE SOCIAL AND HISTORICAL CONTEXT

Agrarian Transformation

Stretching from its southeastern source in the Futa Djalon plateau of Guinea more than 1,700 km westward to its termination at the Senegalese port of St. Louis, the SRB constitutes the third largest wetland ecosystem in Africa (Grosenick et al. 1990:120; Platon 1981:4). The basin covers an estimated 289,000 km^2 and supports a population estimated at 1.6 to 1.7 million inhabitants (Ndiame 1985:3; Van Lavieren and Van Wetten 1990:32). The boundary of the basin also binds the politico-economic ambitions of Senegal, Mali, and Mauritania and has served as the key landmark shaping the course of history and geo-political relations in the region for centuries. The relevance of the Senegal River as a catalyst in the political strife of its neighboring states has recently taken on dramatic proportions. In April 1989, a major conflict erupted between Mauritania and Senegal, in which sovereignty over the river was vehemently disputed.

Geographically, the basin extends across a broad landscape that is demarcated by three distinct regions: the upper basin of humid, tropical forests on the Guinea/Mali border; the middle basin of arid, semi-desert dunes and sandy uplands; and the lower basin of the river delta. Each region varies in terms of morphology, hydrology, and climate, and features a different mix of subsistence activities.

Historically, each geographical zone has also supported distinct ethno-linguistic communities. In particular, the middle valley is inhabited by sedentarized agropastoralists and fishers of diverse ethnic background whose origins date to the fourth century kingdom of Takrur (Boutillier et al. 1962:16). The majority ethnic population, the Haalpulaar,[1] has a long history of political strife, internecine warfare, and conversion to Islam in the eleventh century. By the mid-fifteenth century, the middle valley came under the reign of an aristocratic dynasty of nomadic herders, the Deniyanke FulBe, who established an autonomous geo-political region known as the Fuuta Tooro.

By the sixteenth century, the region reached its zenith in prosperity, producing a surplus of grains, particularly rainfed millet, for inter-regional trade. The sovereignty of the Deniyanke lasted until the late eighteenth century when dominion over the valley's fertile floodplains was usurped by a community of Muslim clerics known as the *ToorodBe* (Boutillier et al. 1962:17; Minvielle 1985:70–71). This period of theocratic jurisdiction marked the institutionalization of Islamic law and the restructuring of customary land-use rights. In particular, pre-Islamic patterns of horizontal land inheritance from brother to brother through the patriline were superseded by vertical lines of land transmission from father to son, a customary Islamic practice still observed in the middle valley today (Minvielle 1985:72).

European contact with the region dates roughly from 1500–1750, during which time Portuguese, Dutch, Spanish, and French companies began trading luxury goods for ivory, gold, salt, gum arabic, leather, and slaves (Boutillier et al. 1962:16). By the late nineteenth century, commercial and political control of the region came under French hegemony through military conquest and the expansion of capitalist markets via extraction of raw materials, particularly gum arabic (Boutillier et al. 1962:18). By 1848, slavery was abolished (Minvielle 1985:170), obliging authorities to institute a policy of taxation through corvée labor. By 1888, a demand for cash was created by enforcing a policy requiring tax payment in currency (Weigel 1982:21). These cash demands coincided with a decline in the gum trade and the establishment of peanut export crop production south of the region. Infrastructural development took place during this time with the newly constructed Dakar-Niger railroad linking the peanut basin and the interior to the coast. Following the decline of the regional economy and the shift of infrastructural resources further south, the need for cash fueled a massive migration out of the river valley. Young men, known as *navetanes*, sought seasonal wages as laborers in the peanut basin, in railroad construction, and in the commercial trade networks along the river (ibid.:72–73). These developments, taken together, seriously eroded the productive labor capacity of the Haalpulaar household and contributed to the growing deficit in domestic food production in the region by the latter half of the twentieth century.

In the post-colonial era since 1960, Senegal has experienced the increasing mobility of labor and the growing penetration of commodity market relations. These features underscore the transitional nature of the post-colonial domestic economy, in which pre-existing social configurations continue to be transformed as a result of partial market integration. A new social constellation of peasant-workers (Murray 1981; Bernal 1991) is emerging in which an ever-expanding reserve of floating migrant labor moves seasonally between rural and urban landscapes.

Corporate Hierarchy

These changes occur in a social context of corporate hierarchy and cultural organization similar to that of other agrarian societies in the West African Sahel. The social structure of Haalpulaar communities is stratified according to criteria of age, gender, and ascribed status at birth. Reminiscent of Meillassoux's (1981) domestic community in which goods, services, and nubile women circulate according to the dictates of an androcentric gerontocracy, a hierarchy of social relations exists among the differentiated segments of Pulaar society – elders and juniors, men and women, patrons and clients – in which control of the means of production has resided historically in the hands of the landed aristocratic class of Muslim clerics, the *ToorodBe*.

The central feature of Haalpulaar society shaping the differential access to resources is the marked distinction among land-rich and land-poor peasants. This distinction is based on hierarchical social relations determined by group social strata, or castes. This attribute of ascribed corporate identity diverges somewhat from the standard age- and gender-based social differentiations generally identified in West African societies (e.g., Meillassoux 1981). Much scholarship on the valley portrays the Pulaar community as a caste society, for lack of a better term to describe the vertical ordering of groups (Boutillier et al. 1962; Wane 1969; Minvielle 1985; Schmitz 1986).

A complex web of patron-client relations characterizes social inter- actions among the various castes. This involves reciprocal and non- reciprocal social obligations in farm production and the concomitant exchange of gifts, goods, and domestic custodial services. Social status and political standing in the community are determined at birth based on one's genealogical position among apical ancestors within the rigid social hierarchy. Thus, the freeborn *rimBe* are situated at the apex of society (Table I). They possess freehold rights to land and exercise the greatest degree of social autonomy and political authority. Control over strategic resources in land and labor has historically resided with the freeborn *rimBe* lineages, particularly those of *ToorodBe* (mara- boutic), FulBe (nomadic), and *SebBe* (warrior) origin. Under *rimBe* patronage are the servile strata of artisans (*ñeeñBe*) and bondservants (*jeyaaBe*), who may be called upon to provide domestic and agricul- tural labor to their freeborn patrons. Many dependents of the politi- cally dominant lineages remain landless today, indebted to the *rimBe* through various share-tenancy and usufruct arrangements.

Haalpulaar social groupings thus feature a substantial degree of sociocultural and occupational diversity, and considerable variation in social composition is found within area communities. Table I lists 24 statutory groups (*HiinDe*) – a rich cultural and occupational mosaic of farmers, herders, fishers, artisans, griots, praise singers, Muslim clerics and scholars, and indentured as well as non-indentured servants.

Social Composition and Land Access in the Study Villages

The degree of social heterogeneity within the populace varies accord- ing to distinct histories at each locale. Variability, not only in social composition but also in land use patterns and strategies of resource differentiation, is illustrated in this study of the three valley communi- ties (Table II). Thus, according to a census taken in late 1988 and early 1989, Dembaka and Thiemping vary in population by a factor of more than ten, with 1,734 inhabitants in Thiemping and only 157 in

Table I. Social Stratification Among the Haalpulaar in the Middle Senegal Valley

Ranking Social Stratum (*HiinDe*)	Occupational Specialization
RIMBE (freeborn)	
ArdiiBe (politico-religious functionaries)	
ToorodBe	spiritual leaders; farmers
FulBe jeeri/waalo	former landed elite; herders
HuuñBe (courtesans)	
SebBe	former warriors; farmers
SabalBe	fishers
JaawamBe	counselors, advisors to ToorodBe
ÑEEÑBE (artisans)	
Fecciram Golle (manual artisans)	
WayluBe BaleeBe	blacksmiths
WayluBe SayakooBe	jewelers
MaabuBe SañooBe	weavers
SakkeeBe WodeeBe	leathermakers (sandals, sacks)
SakkeeBe AlawBe	leathermakers (shoes, harnesses)
LawBe LaaDe	canoe builders
LawBe WorworBe	woodcarvers (domestic utensils)
BuurnaaBe	female potters
ÑaalankooBe/ÑaagotooBe (entertainers, praise singers)	
WammbaaBe	guitarists (*hoddu* players)
MaabuBe Suudu Paate	praise singers of FulBe
MaabuBe JaawamBe	praise singers of JaawamBe
LawBe Gumbala	praise singers of SebBe
AwluBe	griots, historians
JEYAABE (slaves, bondservants)	
RimayBe (emancipated servants)	
MaccuBe SootintooBe	freed slaves by payment
MaccuBe DaccunooBe Allah	freed slaves by Koranic study or battle
MaccuBe TajBeboggi	freed slaves by escape, migration
HalfaaBe (dependent servants)	
MaccuBe HalfaaBe	servants of freeborn and artisans
MaccuBe SañooBe	servant weavers of freeborn
MaccuBe MaccuBe	servants of emancipated maccuBe

Source : adapted from Wane (1969:33).

Table II. Demographic Composition of Tiemping, Thially, and Debaka by Social Stratum, 1989

Ranking Social Stratum	Thiemping		Thially		Dembaka	
	N	(%)	N	(%)	N	(%)
RimBe						
ToorodBe	267	15.4	528	38.9		
SebBe	639	36.9	8	0.6		
FulBe Waalo	107	6.2	104	7.7		
FulBe Jeeri					75	47.8
SubalBe	111	6.4	666	49.1		
ÑeeñBe						
MaabuBe	83	4.8				
WayluBe	92	5.3				
AwluBe	2	0.1				
JeyaaBe						
MaccuBe	412	23.8	34	2.5	82	52.2
Other [a]						
SafalBe	11	0.6	16	1.2		
JolfuBe	10	0.6				
Totals	1734		1356		157	

[a] This category includes four non-Pulaar families residing in Thiemping and Thially. *SafalBe* (s. *Capaato*) refers to Moor or Mauritanian in Pulaar and *JolfuBe* (s. *Jolfo*) are the Wolof (Fagerberg-Diallo 1983:78).

Dembaka. Thially is a more moderate-sized village than Thiemping, with 1,356 residents.

The social composition of Thiemping illustrates the extreme variability in livelihood and social hierarchy found in some Haalpulaar communities. Social groups in Thiemping include, first, the freeborn *rimBe*, comprising almost two-thirds of the village population (64.9%), of which the *SebBe* (former warriors) are the largest group (36.9%); second, the artisan *ñeeñBe*, including two groups of *MaabuBe* weavers and *WayluBe* smiths, constituting 10.2% of the population; and third, a sizeable population of *jeyaaBe* bondservants, making up almost one-quarter (23.8%) of the population.

In contrast to Thiemping, group homogeneity is more the rule than the exception at both Thially and Dembaka. Thially is bifurcated into two residential quarters including Thially Makka, made up almost solely of families of maraboutic origin (*ToorodBe*) engaged in farming and small livestock production, and Thially Subalo, whose residents are of fisher origin (*SubalBe*). Together, the *ToorodBe* clerisy and *SubalBe* fishers constitute the overwhelming majority of the village population (88%), whereas most other groups are absent (e.g., *ñeeñBe*) or nearly so (e.g., *jeyaaBe*, *SebBe*).

Finally, the social composition of Dembaka reflects the strong, if not predominant, association of transhumant pastoralism with one socio-ethnic group, the FulBe *jeeri*. The two herding camps at Dembaka are almost equally divided among the FulBe (47.8%) and their freed slaves (*MaccuBe*, 52.2%).

The contrasts in demographic and social composition at Thiemping, Thially, and Dembaka are also accompanied by differences in agrarian practice. Access to the widest range of farm lands is most evident at Thiemping. This contrasts with the modest resources of the Thially farmers and fishers and the Dembaka FulBe agropastoralists. Land use practice at the three villages sheds light on the key importance of social and resource heterogeneity in the reproduction of thousands of farming, herding, and fishing households in the valley today.

THE ECOLOGICAL CONTEXT: NICHE EXPLOITATION OF THE SENEGAL WETLANDS

The middle Senegal Valley features a web of ecological niches that constitutes a wetland ecosystem. This melange of terrain transects the valley floor beginning at the river, rising above the high levees, descending across the valley floor, and gradually rising again to the high, sandy dunal region distant from the river. Subsistence patterns of farming, herding, and fishing are found within the transect and are practiced in varying combinations according to the features of each microenvironment. These ecological zones are dispersed laterally from the river across the landscape (Figure 3). According to the local classification of land and soils, which is based on water retention capacities (Tabor and Bâ 1987:5), the topographical zonation constituting the middle Senegal wetlands begins with the river itself (*maayo*), then stretches horizontally from the river bank *falo*, to the levee *foonde*, the floodplain or bottomland *waalo*, the upland terrace *jeejegol*, and the adjacent upland *jeeri* at the wetlands periphery.

The ecology of the region is further characterized by a distinct seasonal variation in climate. Precipitation falls intensively during a short two- to three-month period followed by a long, hot dry season. Rains begin in June, increase in intensity and duration by August and September, and abate in October. Annual runoff is distributed in a single peak, with most of the flow concentrated during the truncated

110

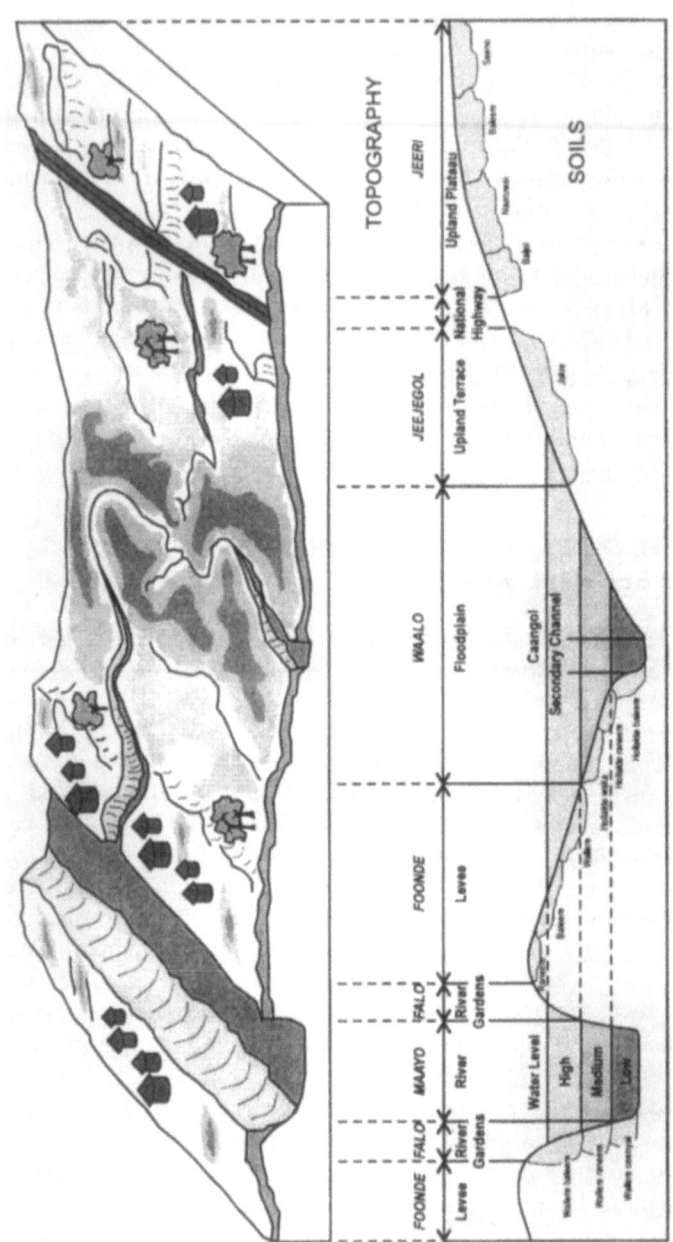

Figure 3. Lateral zonation of the Senegal Valley wetlands.

wet season (Sir Alexander Gibb and Partners et al. 1987). The marked differential in runoff entering the river channels between wet and dry seasons creates a distinct pulse stable ecosystem (Howard-Williams and Thompson 1985:219) in which low-lying basins adjacent to the main river channel are briefly flooded at the height of the rains.

This seasonal flooding of the wetlands enables its occupants to exploit the floodplain intensively, undertaking a series of agropastoral and fishing activities that follows the calendrical ebb and flow of the river's flood cycle. At the beginning of the western African rainy season, runoff from local rainfall allows farmers to plant a variety of sorghum (*fela*) that is well adapted to the dry conditions of the high levee and floodplain *foonde*. This period also corresponds to the planting of finger millet (*suuna*) by farmers and herders dwelling in or near the upland *jeeri* zone. When the river floods, the *waalo* bottomlands also become the site of intensive fishing activity. Soon after the retreat of the floodwaters, intensive recession cultivation begins, first, in the heart of the floodplain, where several varieties of sorghum (*samme*) adapted to varying soil conditions are planted. Second, planting begins on river garden *falo* plots, where inhabitants of riverine villages, particularly older women, practice mixed cropping of vegetables, legumes, grains, and fruit. Third, in years of ample flooding, farmers also exploit high floodplain *foonde* (pl. *poode*) adjacent to the riverine villages. Here, a rich polyculture of grains, vegetables, legumes, and fruits is planted; in years of low flooding, rainfall runoff still permits a sizeable crop to be harvested in the *foonde*.

The succession of farming and fishing activities and the completion of the wetland production cycle are brought full circle during the *waalo* sorghum harvest, when FulBe livestock from the *jeeri* enter the bottomlands to graze the sorghum stubble. This mass migration of FulBe herds into the floodplain, known as the *ñaangal*, permits a rich nutrient exchange of cellulose and manure on the *waalo* lands. The agropastoral dynamic of the floodplain enables the successful reconstitution of area herds during the harsh dry season while simultaneously allowing local farmers and fishers to benefit from enriched floodplain soils.

The seasonality of the wetland regime, however, does not provide a foolproof ecological system, immune to the vicissitudes of climate. Historically, the region has been vulnerable to drought, and rural communities have not always met their subsistence needs by customary practices. Periodic shortfalls in precipitation and nominal flooding of the wetland have occasionally caused food shortage and famine, forcing people to relocate outside the region, as was the case during the Sahelian drought of the late 1960s and early 1970s. In response, rice irrigation schemes were introduced in the middle valley in the mid 1970s by SAED (*Société d'Aménagement et d'Exploitation des Terres du Delta*), a government parastatal, as an agricultural strategy designed to buffer rural peasants from periodic drought. The village-based irrigation perimeters, known as PIVs (*Périmètre Irrigué Villageois*) and GIEs (*Groupement d'Intérêt Economique*), are cooperatively organized by committees of prominent farmers chosen within each community.

Today, valley dwellers engage in a strategy of diversification in response to climatic uncertainty and the ever-growing demands of incorporation within a modern market economy. Their customary repertoire of production activities is augmented by a double-cropping regime of rice pump irrigation in the wet season and irrigated vegetable gardens in the cool dry season (Plates 1–4). Consequently, their use of customary and capital-intensive cultivation, in conjunction with small animal husbandry, fishing, petty commerce, and off-farm employment lessens the chances of scarcity and assures the survival of rural communities.[2]

Wetland Farming and Resource Diversification: The Case of Thiemping

One of the largest wetland expanses in the middle Senegal Valley is found in the Matam region just north of the sub-district administra-

Plate 1. Farming of the *waalo* bottomlands.

Plate 2. Riverine *falo* gardens.

Plate 3. Livestock grazing of post-harvest stover in the *waalo*.

tive center of Kanel (Figure 2). The bottomlands vary in area from less than 1 km² to more than 9 km², depending upon annual flood conditions. The village of Thiemping, located at the northern fringe of the Kanel sub-district on the banks of the Senegal River, has some of the

Plate 4. Preparation of seed beds in a village irrigation perimeter.

most extensive *waalo* bottomland in the area. Smallholders farm more than ten lowland areas, all located in the basin between Thiemping and Kanel. Founding families in the village also formerly owned and farmed productive lands in the *waalo, falo*, and *foonde* across the river in Mauritania, but these were abandoned following the outbreak of hostilities in 1989.

Although *waalo* cultivation is the mainstay of production among Thiemping farmers, the micro-ecology of the region enables additional agricultural activities to be practiced, as well as fishing and animal husbandry. Thus, farmers practice recession agriculture in three distinct zones of the wetlands (*waalo* bottomlands, river *falo*, floodplain *foonde*), rainfed cultivation on the levee *foonde* and upland *jeeri*, and pump-irrigation farming of wet season rice and dry season vegetables. On average, Thiemping households owned or exploited 5.1 fields during the 1989–1990 production season. Of these, 3.3 were recession wetlands (*waalo/falo/foonde*). Although most of the fields were both owned and farmed by the cultivators themselves, a portion were available to be rented out or loaned or farmed in share-tenancy.

Thiemping smallholders thus enjoy a favorable position as farmers of several niches. Nearly all families also possess small herds of sheep and goats, and many own cattle. These herds represent significant holdings for many families – thus the importance of an annually regenerating floodplain habitat of trees, shrubs, and grasses to assure adequate forage.

Furthermore, some households supplement their agropastoral activities by fishing the river channel and the inner floodplain in cycle with the annual rise and fall of the river's waters. Finally, the opening of their *waalo* fields after harvest to herders enhances soil quality and provides a source of nutrient-rich detritus on the plains for the spawning and feeding of migratory fish during flooding. Field access by herders solidifies social and economic ties between the two groups, assuring the continued exchange of grain and milk products.

The capacity of the Senegal wetlands to support such diverse subsistence activities in one village underscores the saliency of diversification as the means of achieving food security by Thiemping smallholders. Families also engage in local petty trade and migrate to cities in search of cash income to assure the ongoing maintenance and reproduction of their households.

Irrigation Farming and Resource Constraint: The Case of Thially

In contrast to the mix of agropastoral activities at Thiemping, the neighboring village of Thially (Figure 2) is a smaller farming and fishing settlement with distinctively different demographic and occupational features as well as a more uniform resource base. Producers at Thially Makka engage primarily in irrigated rice farming and carry out some supplementary small livestock production. Residents of Thially Subalo farm irrigated perimeters and fish the river and interior basin areas of the nearby Mauritanian floodplains. All Thially inhabitants, however, experienced the sudden loss of rich farm lands and fertile fishing grounds on the Mauritanian right bank after the expropriation of these areas by Mauritanian *bidan* and *haratin*[3] civilians during the border conflict that broke out in April 1989.

Before the onset of hostilities between Senegal and Mauritania, the riverine lands in Mauritania, particularly large fertile expanses of river garden *falo*, were key in provisioning the village well into the dry season when other food stocks were depleted. In years of ample flooding, the surface area of these Mauritanian fields approached that usually found only in the *waalo* bottomlands,[4] and during the period of flood recession, several village families would relocate each year to the Mauritanian side. The magnitude of these lands lost to Thially farmers is indicated in Table III. Compared to 1988, land use in 1989

Table III. Loss of Thially Transborder Landholdings, 1988–1989

Production System	Fields Cultivated in 1988[a]	Fields Cultivated in 1989	Percentage of Fields Lost in 1989
Waalo	46	6	87.0
Falo/Foonde	159	13	91.8
Jeeri	120	68	43.3
Total	325	87	73.2

[a]Total number of fields owned, farmed, or rented out (Senegal and Mauritania).

was reduced by 87 percent of *waalo* and nearly 92 percent of *falo/foonde* lands. The percentage of rainfed *jeeri* lost in 1989 was about one-half the loss sustained in the recession systems. In total, nearly 91 percent of *waalo* and *falo/foonde* lands were not farmed following the border conflict, forcing Thially residents to hastily adapt to land scarcity by investing all available resources of labor and capital in pump-irrigated rice cultivation. On average, Thially producers farmed 3.1 fields during 1989–1990, and unlike the land use profiles in Thiemping, recession lands (0.1 *waalo* and 0.1 *falo/foonde* fields) were a meager component of their holdings. For many, this change was debilitating, severely constraining labor mobility and resulting in a modest crop output incommensurate with the relatively high investments in human and capital resources.

Furthermore, the most important fishing grounds for Thially families had been on the Mauritanian side of the river, several kilometers to the interior, where there is a fertile floodplain and large residual pond-marsh area known as Mehi. Thially fishers, as well as many others from throughout the region, exploited this fishing ground throughout the year, claiming it to be one of the richest fishing grounds in the middle valley. At the height of the flood cycle, this area was fished by canoe for several weeks until withdrawal of the floodwaters. In the dry season, the large bodies of water provided small fish catches for daily consumption at a time of year when local diets were low in protein.

As a result of the 1989 conflict, as well as rain shortfalls and poor flooding in recent years, Thially fishers have been forced to abandon this fishing ground and other areas along the river. If not replaced, the loss of these areas could mark the demise of fishing as a viable livelihood for villagers. In the meantime, family strategies of social reproduction have been narrowed, compelling individuals to make hard, new choices between on- and off-farm subsistence and wage-earning activities. Many families have elected to intensify their efforts in irrigation agriculture, placing new demands on already strained family budgets and limited supplies of labor.

Seasonal Herding of the Bottomlands:
The Case of Dembaka

The sandy dunes of the upland *jeeri* plateau bordering the wetland are home to thousands of FulBe herders. For centuries, this semi-nomadic population has been making annual treks northward into the valley floodplain in search of green pasture. June rains in the SRB bring the northern drive of thousands of cattle, sheep, and goats from southern pastures in Senegal, Guinea, and Mali. Between June and September, as the dry season approaches and the southern pasture becomes depleted, herds from the southern *jeeri* move slowly northward toward the plains. Finally, as rains subside by mid-September, floodplain farmers begin planting their *waalo* fields. When the *waalo* harvest nears in March, herds forage along transhumant routes near the floodplain following a corridor from the uplands to the river, where small pond and marsh areas (*beeli*) serve as critical water points.

Benefiting from their early historical control of the floodplain for grazing purposes, many FulBe communities continue to possess sizeable *waalo* landholdings (Niasse 1989). Thus, of the 3.9 fields farmed on average by Dembaka producers, 2.2, or more than one-half, are *waalo* lands. However, landholding status within any given village may be highly differentiated among individual families. Some FulBe without ownership rights in the *waalo* migrate seasonally from the *jeeri* to the floodplain to set up temporary camp (*seedaanooji*) and farm both rainfed *foonde* and recession *waalo* fields as tenant farmers to landowners near the river. They are accompanied by family herds that pasture in the uncultivated areas until harvest fields are accessible. Ruminants experience a significant loss in body weight during the long dry months. Thus access to *waalo* pasture becomes critical to the reproductive survival of large numbers of area herds. In a good flood year such as 1988, *waalo* crop residues are able to sustain large numbers of livestock for up to one month. Dembaka herders are unable to access *falo/foonde* and irrigated lands. The loss of bottomlands in the absence of sustained flooding would clearly place excessive hardship on these herders to seek alternate paths to meet the nutritional needs of their families and livestock.

At Dembaka, dry season production activities culminate in the communal grazing (*ñaangal*) of the Thiemping bottomlands in March. This seasonal activity brings the FulBe herding community into closer contact with riverine farmers and fishers, promoting the exchange of grain, fish, and dairy products. In particular, FulBe women sell or exchange surplus milk in local markets such as Thiemping, generating critically needed revenues for the weekly purchase of foodstuffs. This period also corresponds to the natural chronology of flood drawdown in each basin. In this farming complex, crop harvests occur in a consecutive rather than overlapping fashion, and farmers cultivating in more than one bottomland benefit from the staggering of sorghum harvests as they move from one field to the next. Herders also profit from the customary practice of extended access to the fields after harvest. At Dembaka they follow the harvest schedule intently, pasturing their animals from two to four days in each bottomland until crop residues are well depleted. They then migrate to a contiguous area that has just opened for grazing. As the harvest proceeds and each *waalo* bottomland is grazed and manured, the cycle of nutrient exchange between farm and herd is completed. By mid-April, late in the dry season, herders return south to encampments and pastures in the upland *jeeri* plateau.

FOOD SECURITY AND RESOURCE ALLOCATION: AN INTER-VILLAGE COMPARISON

The ability of middle valley farmers to meet their food needs through on-farm production depends in large part on the resource configuration of the household. Two factors intrinsic to farm production – land and labor – are key in determining relative levels of food security. The research findings in this section demonstrate quantitatively that a broad array of farming activities, both customary and capital-intensive, provides a greater level of food security to the household than reliance on one principal form of agricultural exploitation. Moreover, the broad-based strategy of multiple cropping at Thiemping provides substantially greater returns per unit labor to the producer than heavy investment of limited resources in monocultural strategies such as the capital-intensive pump-irrigation of rice as practiced at Thially.[5] Irrigated rice farming, on the other hand, has frequently been extolled by development planners in the region for its capacity to provide superior crop yields. Data on crop harvests presented in this section confirm this observation. However, improved productivity (in terms of returns per unit land) comes at the expense of a disproportionate investment of labor and capital.[6] Finally, comparative analysis of food entering households on a monthly basis shows that per capita levels of grain and vegetable output (yields/capita) at Thiemping are superior to those at Thially. Thus, stock flows under regimes of multiple resource management provide a constant output per capita, whereas harvests under conditions of irrigated rice monoculture peak unimodally in one month, followed by a long period of food scarcity during the remainder of the year.

Returns Per Unit Land and Labor

Households in the Thiemping sample, on average, cultivated five fields during the 1989–1990 season (Table IV), whereas producers in Thially farmed only three fields. This greater number of fields farmed is the result of access to fertile *waalo* bottomlands (two fields per household) as well as other agro-ecological zones.[7] Due to the loss of bottomlands in Mauritania, recession farming in Thially (*waalo, falo, foonde* systems) is rare (0.1 fields). In order to compensate for the loss of highly valued lands, Thially producers were forced to depend heavily on their irrigation scheme (PIV), farming more than one plot (1.6) per household. A scarce reserve of labor and capital in the village also led many producers to farm millet under rainfed conditions (*jeeri*), averaging 1.3 plots per family.

Table IV. Comparison of Field Plots Farmed, 1989–1990

Production System	Households In Sample		Field Plots Farmed Per Household (avg)		Average Farmed Per Household (ha)	
	Thiemping	Thially	Thiemping	Thially[b]	Thiemping	Thially
Waalo	14	1	1.9	0.1	4.64	0.21
Falo	9	1	0.8	0.1	0.14	0.00
Foonde	5	0	0.6	0.0	0.13	0.00
Jeeri	13	8	1.3	1.3	0.50	0.49
Garden	8	0	0.5	0.0	0.02	0.00
GIE/PIV	1	7	0.1	1.6	0.26	0.50
Total	16[a]	8[a]	5.1	3.1	5.68	1.20

[a] Total number of households in the sample.
[b] Number of field plots farmed after loss of land to Mauritania in April 1989.

Figures on surface area farmed per household (Table IV, columns 6 and 7) show even greater disparities between the villages. Thus, Thiemping has an advantage over Thially in the average area farmed per household by a margin of nearly 5 to 1. This is largely the result of superior floodplain (*waalo*) field size, totalling nearly 5 ha per household. Thially residents, in contrast, have no fields exceeding one-half ha (*jeeri*, PIV).

In Table V, labor time allocation and crop harvest data gathered at Thiemping and Thially illustrate the advantages and disadvantages of each cropping system. Thiemping cultivators, on average, deploy nominal levels of labor (61 person-days/ha) despite the wide variance of labor input among the various systems. The highly labor-intensive nature of irrigation systems (771 person-days in gardens, 196 person-days in the GIE) is in contrast to the modest levels of labor time invested in the recession systems (32 in *waalo*, 95 in *falo*, 126 in *foonde*).[8] Conversely,[9] Thially farmers invest five times more labor per hectare than those in Thiemping, averaging 310 person-days.

By ignoring the additional resources of labor and capital that critically shape its outcome, irrigation farming in the middle valley is all too often lauded, in comparison to other production regimes, for its relatively high performance. Yet when output is defined in relation to the amount of labor invested, returns no longer seem high. A comparison of production figures from Table V (gross output/person-day) illustrates the appeal of customary recession farming when labor time is balanced against crop output. At Thiemping, the highest levels of crop output for each unit of labor clearly fall in favor of *waalo* and *falo* cultivation (12.6 and 13.5 kg/person-day respectively). In terms of

Table V. Comparison of Labor Input and Crop Output, 1989–1990

Production System	Labor Input (person-days/ha)		Crop Output (kg/ha)		Gross Output/Person (kg/person-day)	
	Thiemping	Thially	Thiemping	Thially	Thiemping	Thially
Waalo	32	278	403	91	12.6	0.3
Falo	95	693	1,282		13.5	
Foonde	126		456		3.6	
Jeeri	192	107	186	357	1.0	3.3
Garden	771		6,833		8.9	
GIE/PIV	196	517	1,128	3,334	5.8	6.4
Average	61	310	470	1,548	7.3	2.5

crop output alone, Thially producers have a clear advantage. Harvests per unit area farmed average nearly four times (1,548 kg/ha) those of Thiemping (470 kg/ha). But when crop output is measured against labor input, the ratio of Thiemping (7.3) exceeds that of Thially (2.5) by a margin of nearly 3 to 1.

Although high productivity is apparent in the irrigation regimes of both villages, yields in rainfed and recession farming are much lower. Mixed crop production is relatively high in the *falo* gardens of Thiemping (1,282 kg/ha), as a polyculture of grains, legumes, and cucurbits provides a varied supply of nutrients to the local diet and is a vital source of household income for women farmers well into the dry season. The important role of gardens has gone virtually unnoticed by development planners in the region. A pro-irrigation policy of flood suppression would severely threaten the ecological viability of gardening and have detrimental consequences for the economic status of women who rely on the gardens to generate small revenues.

These data underscore the fact that smallholding peasants in the region are more concerned with mitigating risk and minimizing labor than with maximizing production. When possible, producers such as those in Thiemping choose to free up labor for a broad range of production options,[10] whereas those in Thially are forced to intensify most of their family labor and capital in rice irrigation. The degree of risk in undertaking such an intensive mono-cropping strategy is particularly high in a precarious Sahelian environment where labor and capital are frequently in short supply.

The Seasonality of Food Security

A wetland habitat provides a sequential spacing of crop harvests over the course of a year, thereby reducing cyclical phases of crop surplus and shortage between wet and dry seasons and attenuating bouts of food scarcity. Characteristic of the Sahel, a sharp flux in seasonal cropping patterns results in wide variation in the month-to-month level of food supply of small farm communities. Hence, a comparison of levels of food security that derive exclusively from farm production at Thiemping and Thially indicates the capacity of each production system to provision the staple needs of the household throughout the year.

Data on harvests (Table VI) reveal the relative contribution of each farm activity to household food supply. Sorghum harvests from the Thiemping floodplain (*waalo*) provide the largest crop per capita

Table VI. Comparison of Per Capita Crop Harvests by Production System, 1989–1990

Production System	Thiemping		Thially	
	Grains (kg)	Vegetables (kg)	Grains (kg)	Vegetables (kg)
Waalo	231.7		2.1	
Falo	8.7	6.6		
Foonde	1.6	1.1		
Jeeri	9.3		14.5	0.3
GIE/PIV	7.7	29.1	190.5	
Garden		22.3		
Total	258.9	59.1	207.1	0.3

(231.7 kg), followed by irrigated rice (PIV) at Thially (190.5 kg). The broader array of cropping systems at Thiemping also provides residents with a varied diet of vegetables and fruits that is particularly important during the long dry season when granary stocks are low. This abundant mix of horticultural crops was conspicuously absent from the diet of Thially households in 1989–1990 and necessitated greater cash expenditures to augment the local diet with vegetable purchases in the village market. Overall, figures in Table VI verify that Thiemping farmers achieve a greater level of food security from on-farm production than those at Thially.

A staggered sequencing of agricultural tasks enables Thiemping peasants to farm the *waalo*, *falo*, and *foonde* with few critical labor bottlenecks during peak periods of the farming season. *Falo* and *foonde* gardens act as a relay cropping system (Adams 1983:295), providing vegetables and cereal grains for human consumption and fodder for small livestock between the rainfed *jeeri* harvest in November and the *waalo* harvest in March. Since the early 1970s, this succession of cropping systems has been followed by an irrigated crop of garden vegetables that supplements local diets at a time when granary reserves are depleted. The staggering of *falo* and *foonde* harvests, particularly in the absence of a rainfed crop, acts as a food reserve, buffering against shortages well into the dry season until harvest of the *waalo* sorghum crop.

This cropping sequence provides modest but steady inflows of cereals and vegetables throughout much of the year (Table VII). Thus, at Thiemping, there is a total absence of a harvest in only three months

Table VII. Comparison of Per Capita Crop Harvests by Month, 1989–1990

Month	Thiemping			Thially		
	Grains (kg)	Vegetables (kg)	Total (kg)	Grain (kg)	Vegetable (kg)	Total (kg)
January	1.1	5.5	6.6			
February	1.7	4.7	6.4	2.1		2.1
March	233.0	1.8	234.8			
April	2.3	1.6	3.8			
May						
June						
July						
August	7.7	2.2	9.8			
September		3.9	3.9			
October	1.7	0.6	2.3			
November	10.0	30.0	40.0	14.5	0.3	14.8
December	1.4	8.8	10.2	190.5	0.0	190.5
Total	258.9	59.1	317.8	207.1	0.3	207.4

of the dry season (May–July); however, food stocks rise at the end of the wet season in November with the harvesting of rainfed millet in the *jeeri* and garden crops in the *falo*, *foonde*, and irrigated perimeter (GIE). Grain reserves peak in March with a sizable *waalo* sorghum harvest (233 kg/capita) that is then slowly depleted during the lean months of the dry season. The most vulnerable period usually extends from August to September, when the sorghum stock has been reduced and farmers await the harvest of their rainfed and garden crops in October and November.

In contrast to Thiemping, Thially's sequential harvests are less successful. Here, cultivators must rely on the rice harvest in December (190.5 kg/capita) to carry them through the rest of the year. Given the absence of other crops, Thially farmers become highly susceptible to periods of food scarcity.[11] The figures in Table VII illustrate the degree of risk that may be associated with a future river basin management policy that calls for suppression of the river's natural flood cycle. Under such a scenario, farmers in the valley would be forced to rely on only one or possibly two principal crop harvests throughout the year. Proponents of full-scale adoption of irrigation agriculture propose a second cropping season during the long dry months from January to June. Under such conditions, a second cereal crop would be harvested at mid-year, buffering the rapid depletion of grain stocks during the lean period of the year. A caveat, however, lies in the ability of farm households that are already hard pressed with chronic shortages of cash and labor to mobilize additional levels of these resources, particularly during a phase of the agricultural calendar when many households prefer to free up labor for wage earning opportunities off the farm.

THE DEVELOPMENT DISCOURSE: DEFENSE OR DEMISE OF THE SENEGAL WETLANDS

Data collected at Thiemping, Thially, and Dembaka show that middle valley producers optimize their food security when the broadest range of options are available to them. Thus, smallholders such as those in Thially, relying on monocultural strategies, have fared poorly in providing food for their families. I argue that production data from these settings are predictive of future food insecurity problems for peasants compelled to rely exclusively on irrigated farming in a post-dam scenario of flood cessation. Although the constellation of landholdings at Thially may be atypical of resource profiles for many basin communities, these examples nevertheless reveal the vulnerability of valley smallholders when land is taken out of production, whether by forcible expropriation or by loss of flooding due to river impoundment. These settlements are harbingers of the food insecurity problems that may threaten the region in the years to come. By contrast, however, Thiemping farmers are more secure in achieving superior levels of on-farm per-capita food production. These findings call for a reevaluation of policy that advances the expansion of irrigation agriculture, while foregoing the preservation and enhancement of customary production activities.

This chapter focuses on resource endowments and shows why peasant farm communities in the middle Senegal Valley are better able to attain levels of food security through a diverse mix of livelihoods than by rigid adherence to rice mono-cropping. Farming, herding, and fishing communities in the middle Senegal Valley have co-evolved with one another and have sustainably exploited a fertile seasonal wetland for centuries, but the renewal of the ecosystem and the welfare of the inhabitants who rely upon it for their survival is increasingly undermined by a faltering Sahelian environment. Little attention has been given to the self-regulating dynamics of the wetland ecosystem, and to enhancement rather than dismemberment of the natural habitat and the existing production systems that accompany it.

Many endangered wetland ecosystems on the African continent such as the floodplains of the middle Senegal River Valley have been severely undervalued for their capacity to sustain human communities and associated flora and fauna. Until recently, mainstream models of river basin development have reflected little concern for the environmental and socio-economic consequences of dam impoundment on those populations most directly affected by the alterations of Africa's large rivers – that is, small, isolated agrarian wetland communities. This chapter suggests that the livelihoods of more than one million valley inhabitants are seriously endangered by the planning stratagems of river basin policy makers and development officials.

In the policy context of post-dam development of the Senegal River Basin – in which two costly dams, the Diama and the Manantali, have been constructed as the cure-all to the region's economic ills – the decision to terminate flooding of the middle valley plains will have significant deleterious consequences for the communities discussed in this chapter. In farming communities that depend heavily on customary recessional regimes (*waalo*, *falo*, *foonde*) such as Thiemping and Thially, the loss of fertile lands and the consequent shift to irrigation farming could have the paradoxical effect of *accelerating* rather than *impeding* migration outside the region. Recent research findings from the middle Senegal Valley suggest that adult men may increasingly opt to seek off-farm wages in order to finance the rising capital costs of production in village rice schemes (Diemer and van der Laan 1987:93,104; Niasse, in Horowitz et al. 1991:223). This observation has, at least until recently, been counter to the policy objective of the tri-state river basin management authority, OMVS, which has been intent on retaining rural farm labor through the expansion of irrigation in the river basin.[12] Seasonal herding communities located at the floodplain periphery, such as the FulBe agropastoralists at Dembaka, have been largely overlooked in the development optic of river basin planners as well. Failure to consider the integral relationship between semi-nomadic communities and their reliance on recessional lands for seasonal cropping and herding could spell the displacement of these seasonal settlements. This, in turn, would have serious dislocative effects on the economy of neighboring riverine villages like Thiemping as well.

In recent years there has been mounting environmental opposition to the planning stratagems of river basin development technicians and

dam engineers in an international context. Some environmental con-
stituencies have been calling for a moratorium on the future construction of high dams in fragile ecological settings such as the Sahel. Until
recently, many international donors and civil engineering contractors
held firm in their agenda to proceed with business as usual – that is, to
re-engineer riverine habitats in a manner that contradicts international treaties on biodiversity conservation and turns a deaf ear on the
needs of the most poor.[13]

Alternative solutions to the conservation-river basin development
impasse have recently been proposed calling for a dam-enhanced
simulation of natural flood conditions that will assure downstream
flooding of low-lying basin areas and secure the continued productivity of customary farming, fishing, and livestock regimes (Scudder
1989; Horowitz et al. 1991; Magistro 1994; Salem-Murdock et al.
1994). This proposal embraces a multifunctional approach to resource
development in river basins, in which demands for hydroelectricity do
not necessarily preclude the preservation and amelioration of downstream production regimes. A revision of policy on river basin development will necessitate further anthropological research on flood-
dependent systems of farming, range production, and fisheries that
validates or refutes the preservation of such regimes on both environmental and economic grounds. In this vein, Scudder calls for cost-
benefit analyses that will either justify or dismiss such a course of
action. He supports a revised conceptualization of river basin management that operationalizes the current rhetoric of "sustainable development" by factoring in the value of the local resource base, both human
and environmental:

> For development to be sustainable, conservation, local participation,
> and poverty alleviation must be linked. For river basin development,
> this means that current accounting of costs and benefits at the national
> level must be complemented by regional, local, and environmental
> accounting. (Scudder 1989: 7)

The argument here linking "regional, local, and environmental
accounting" with "current accounting of costs and benefits" reflects
current discourse among social scientists, ecologists, development
practitioners, and policy analysts concerning the need for a new paradigm of "environmental economics." Such an approach challenges
econometric models of domestic productivity and suggests that a
range of environmental and social variables be fused with standard
measures of domestic growth and national income accounts (Repetto

1992; Cobb et al. 1995). In essence, incremental gains in economic efficiency and productivity must be weighed against real losses to the environment and potential negative consequences for society. This approach is consonant with the underpinnings of the new sustainable development canon within the international foreign aid community.

The social and environmental costs to be borne by the physical reconfiguration of the northern Senegal wetlands have been touched upon indirectly in this study. Further accounting of the tradeoffs between increased agricultural output via rice pump-irrigation at the cost of biotic losses to the wetland ecosystem and disruption of agrarian livelihoods is needed. Research findings call for more precise social and environmental accounting to be factored into the equation of economic productivity in the Senegal River Valley. A fragile semi-arid wetland habitat and the food security of approximately one million inhabitants in the region is threatened by a modern dam infrastructure that may homogenize a diverse biophysical landscape and fragment a cohesive intercommunity nexus of social, economic, and agro-ecological institutions and processes. An informed ecology of food production systems in northern Senegal has been illustrated in this study. The future well-being of smallholding communities in the middle Senegal Valley must derive from an economic and environmental logic that sustains downriver production assemblages and perceives them to be *compatible, not competitive*, with demands for hydroelectric power generation and modern irrigation agriculture. This can be achieved by introducing a more holistic knowledge base of the Senegalese wetlands, in which social and environmental concerns are given more proportional balance in weighing the economic imperatives of river basin development.

ACKNOWLEDGMENTS

I am indebted to the Institute for Development Anthropology (IDA) in Binghamton, New York, under whose auspices field research was carried out from 1988 to 1990 as part of the Senegal River Basin Monitoring Activity, Phase I (SRBMA I). SRBMA I was funded by US AID/Dakar as a buy-in to the Cooperative Agreement on Settlement and Resource Systems Analysis. I am grateful to the Social Science Research Council for dissertation write-up support under an Africa Program Dissertation Fellowship Research Grant.

I am also grateful to Michael Horowitz, my dissertation advisor and co-director of the Institute, and mentors Peter Little, Michael Painter, and Muneera Salem-Murdock, all of whom assisted me during my field research and provided critical commentaries on portions of this paper. Thayer Scudder, co-director at IDA, has also been instrumental in the genesis of this work. Finally, I thank Endre Nyerges and Barbara Cellarius for their adept and prompt editorial review of my work and Vivian Carlip at the Institute for her critical scrutiny of my economic data. Factual errors and observations in this work remain solely mine.

END NOTES

1. The term Haalpulaar ("speakers of Pulaar") refers to the sedentarized agropastoral and fishing population inhabiting the middle Senegal River Valley. Other terms used in the colonial and post-colonial literature in referring to this ethnic group are Toucouleur (French) and Tukulor (English). The terms Haalpulaar and Pulaar are used interchangeably throughout this work.

2. Several previous studies (Boutillier et al. 1962; Lericollais and Diallo 1980; Minvielle 1985; and Schmitz 1986) have highlighted the dynamics of wetland production in relation to the socioecological mechanisms enabling multiple uses of the floodplain.

3. These terms are used to describe a patron class of Mauritanians of Arabo-Berber descent and their subordinate clients of sub-Saharan African descent. *Bidan* and *haratin* communities are found throughout the Senegal River Valley residing in small settlements on the right bank of the river.

4. During the 1988-1989 *falo* growing season, Thially fishing families were visited on the right bank just months prior to the outbreak of hostilities. I had no opportunity to measure these fields but was struck by their size, approaching 0.5-1.0 ha, and the large quantities of maize, squash, and melons awaiting harvest. Unfortunately, the area was abandoned soon after, and families were unable to recover the large harvests belonging to them.

5. This section is based on field data that were collected by means of bi-weekly structured and non-structured surveys and interviews, over a period of 21 months between 1988-1990. The survey population is a random stratified sample of 15 households at Thiemping, 8 at Thially, and 1 at Dembaka, chosen from a broader demographic study of 152 households at Thiemping, 116 at Thially, and 7 at Dembaka. Field sizes were measured and, for each sample household at each interval, labor time allocation and crop outputs were measured and recorded. In order to reduce numbers for comparative purposes, and because of the close association of Dembaka herders with the Thiemping bottomlands, data on the seasonal encampment have been incorporated into the larger Thiemping data set.

6. For comparisons of capital returns in recession and irrigation farming, see Horowitz and Salem-Murdock (1990), Horowitz et al. (1991), Magistro (1994).

7. The exception is in the cultivation of irrigated perimeters (GIEs, PIVs). One household in the Thiemping sample farms the only private perimeter in the village, while eight households participate in a cooperative irrigated women's garden. Thially farmers, on the other hand, depend heavily on access to multiple plots in their community rice perimeter.

8. Person-days are recorded here as a five-hour working day, a realistic estimate of the time spent by farmers in their fields. Age differences in the various tasks are not considered and are not believed to be significant. The exceptionally high figure of 771 person-days for horticulture reflects the laboriousness of attending to crops that require frequent watering, weeding, and surveillance. The high figure of 192 person-days in dryland (*jeeri*) cultivation reflects the vulnerability of this system to erratic rainfall, which may compel farmers to sow and weed fields more than once.

9. The contrast in labor time between the Thially PIV and Thiemping GIE is the result of different crops grown (vegetables versus rice) and varying strategies of labor mobilization. The GIE relies primarily on waged day labor, thus the lower number of hours invested, while the PIV relies more on household labor. The high figures for *waalo* and *falo* farming may reflect the particular organization of the sole sample household farming these systems.

10. For example, communal rice irrigation in Thiemping was abandoned during the 1989–1990 wet season due to extensive inundation of a bottomland area that is rarely flooded and was therefore earmarked for rice cultivation by the village. With the arrival of exceptional floodwaters in 1989, members abandoned rice production in favor of recession farming of sorghum in the irrigated perimeter. This incident and similar instances in the nearby region demonstrates the clear preference of middle valley peasants for low-input cropping systems over higher yielding capital- and-labor intensive agricultural practices such as rice irrigation.

11. Young men have responded by migrating in search of waged income, and migratory remittances are important for household food provisioning strategies. For an assessment, see Horowitz et al. (1991).

12. At the time of writing, the policy position of the OMVS is being re-evaluated. Following the lead of the Senegalese government, donors such as US AID and the World Bank are now considering a policy of controlled floods from Manantali that would preserve the downstream wetlands and attendant mix of customary production activities. If a policy of flood simulation is instituted, it would establish a new precedent for river basin development in Africa to enhance the regional ecology and restore downstream production.

13. The growing dissent of the conservation movement, determined to halt
future large-scale dam projects in Africa and elsewhere, has been discussed
by Scudder (1989:7). Particularly in this discussion, developers of river
basins are portrayed as the antagonists who victimize rather than aid
rural communities by excluding them from the decision-making process
and diverting the benefits of modern dams to the urban rather than rural
populace (ibid.:8; Scudder 1980:392). Scudder (1989:7) notes that the dia-
logue became more combative in the 1980s as river basin planners continued to neglect the needs of those most directly affected by dam
building.

REFERENCES

Adams, William M.
  1983   Downstream Impact of River Control, Sokoto Valley, Nigeria. Un-
         published Ph.D. Dissertation, University of Cambridge.
  1986   Traditional Agriculture and Water Use in the Sokoto Valley,
         Nigeria. The Geographical Journal 152(1): 30–42.
  1987   Approaches to Water Resource Development, Sokoto Valley,
         Nigeria: The Problem of Sustainability. *In* Conservation in Africa:
         People, Policies and Practice. D. M. Anderson and R. H. Grove, eds.
         Pp. 307–325. Cambridge: Cambridge University.

Altieri, Miguel A.
  1987   Agroecology: The Scientific Basis of Alternative Agriculture. Boulder
         CO: Westview.

Altieri, Miguel A., and Susanna B. Hecht
  1990   Agroecology and Small Farm Development. Boca Raton FL: RC
         Press.

Bernal, Victoria
  1991   Cultivating Workers: Peasants and Capitalism in a Sudanese Vill-
         age. New York: Columbia University.

Boutillier, Jean-Louis, P. Cantrelle, J. Causse, C. Laurent, and T. N'Doye
  1962   La Moyenne Vallée du Sénégal: Etude Socio-économique. Paris:
         INSEE.

Brokensha, David, Dennis M. Warren, and Oswald Werner
  1980   Indigenous Knowledge Systems and Development. Lanham MD:
         University Press of America.

Cernea, Michael
  1985   Sociological Knowledge for Development Projects. *In* Putting
         People First. M. Cernea, ed. Pp. 3–21. New York: Oxford Univer-
         sity for the World Bank.

Chambers, Robert
  1983   Rural Development: Putting the Last First. London: Longman.

Cobb, Clifford, Ted Halstead, and Johnathan Rowe
   1995   If the GDP is Up, Why is America Down? Atlantic Monthly 276(4): 59–78.

Commander, Simon, Ousseynou Ndoye, and Ismael Ouedraogo
   1989   Senegal: 1979–88. *In* Structural Adjustment and Agriculture: Theory and Practice in Africa and Latin America. S. Commander, ed. Pp. 144–173. London: ODI.

Diemer, Geert, and Ellen van der Laan
   1987   L'Irrigation au Sahel: La Crise des Périmètres Irrigués et la Voie Haalpulaar. Paris: Editions Karthala.

Duruflé, Gilles
   1988   L'Ajustement Structurel en Afrique (Sénégal, Côte d'Ivoire, Madagascar). Paris: Editions Karthala.

Dyson-Hudson, Neville
   1985   Pastoral Production Systems and Livestock Development Projects: An East African Perspective. *In* Putting People First. M. Cernea, ed. Pp. 157–186. New York: Oxford University for the World Bank.

Fagerberg-Diallo, Sonja
   1983   Advanced Readings in Pulaar. Dakar: American Lutheran Church in Senegal.

GERSAR-CACG, Euroconsult, Sir Alexander Gibb & Partners, SONED-Afrique
   1990   Plan Directeur de Développement Intégré pour la Rive Gauche de la Vallée du Fleuve Sénégal. Rapport et Annexes (document provisoire). New York: PNUD/BIRD.

Grosenick, Gerold, Abdoulay Djegal, Jack W. King et al.
   1990   Analyse pour la Gestion des Ressources Naturelles du Sénégal: Rapport Final. (PDC-5517-I-13-7136-00). Washington, DC: Louis Berger International/Institute for Development Anthropology.

Horowitz, Michael M
   1985   Ideology, Policy and Praxis in Pastoral Livestock Development. *In* The Anthropology of Rural Development in West Africa. M. Horowitz and T. Painter, eds. Pp. 249–272. Boulder CO: Westview.

   1989   Victims of Development. Development Anthropology Network 7(2):1–8.

Horowitz, Michael M, and Muneera Salem-Murdock
   1990   The Senegal River Basin Monitoring Activity: A Phase One Synthesis Report. IDA Working Paper No. 54. Binghamton NY: Institute for Development Anthropology.

Horowitz, Michael M, Muneera Salem-Murdock, Curt Grimm et al.
   1991   The Senegal River Basin Monitoring Activity, Phase I: Final Report. Binghamton NY: Institute for Development Anthropology.

Howard-Williams, Clive, and Keith Thompson
    1985 The Conservation and Management of African Wetlands. *In* The
          Ecology and Management of African Wetland Vegetation. Patrick
          Denny, ed. Pp. 203–230. Dordrecht: Dr. W. Junk.

Lericollais, André, and Yveline Diallo
    1980 Peuplement et Cultures de Saison Sèche dans la Vallée du Sénégal. 7
          Cartes et Notices. Dakar: ORSTOM.

Little, Peter
    1984 Critical Socio-Economic Variables in African Pastoral Livestock
          Development: Toward a Comparative Framework. *In* Livestock De-
          velopment in Subsaharan Africa: Constraints, Prospects, Policy. J.
          Simpson and P. Evangelou, eds. Pp. 201–214. Boulder CO: West-
          view.

Little, Peter, and David Brokensha
    1985 Local Institutions, Tenure and Resource Management in East Afri-
          ca. Binghamton NY: Institute for Development Anthropology.

Magistro, John V.
    1994 Ecology and Production in the Middle Senegal Valley Wetlands.
          Unpublished Ph.D. Dissertation, Binghamton University.

Meillassoux, Claude
    1981 Maidens, Meal and Money. Cambridge: Cambridge University.

Minvielle, Jean-Paul
    1985 Paysans Migrants du Fouta Toro: La Vallée du Sénégal. Paris: OR-
          STOM.

Murray, Colin
    1981 Families Divided: The Impact of Migrant Labour in Lesotho. Cam-
          bridge: Cambridge University.

Ndiame, Fadel
    1985 A Comparative Analysis of Alternative Irrigation Schemes and the
          Objective of Food Security: The Case of the Fleuve Region in
          Senegal. M.S. Thesis, Michigan State University.

Niasse, Madiodio
    1989 September Troisième Rapport: Suivi des Systemes de Production
          Agricole à Doumga Rindiaw, Première Partie: P.I.V. et Waalo. Bin-
          ghamton NY: Institute for Development Anthropology.

Platon, Pierre
    1981 OMVS: The Development of the Senegal River (Special Issue in
          English). Marchés Tropicaux et Méditerranéens (17 April 1981).

Repetto, Robert
    1992 Earth in the Balance Sheet: Incorporating Natural Resources in Na-
          tional Income Accounts. Environment 34(7): 13–20, 43–45.

Richards, Paul
1985   Indigenous Agricultural Revolution: Ecology and Food Production in West Africa. London: Longman.

Salem-Murdock, Muneera, Madiodio Niasse, John Magistro, et al.
1994   Les Barrages de la Controverse: Le Cas de la Vallée du Fleuve Sénégal. Paris: Editions L'Harmattan.

Schmitz, Jean
1986   L'État Géomètre: Les Leydi des Peul du Fuuta Tooro (Sénégal) et du Maasina (Mali). Cahiers d'Etudes Africaines XXVI(3):349–394.

Scudder, Thayer
1980   River Basin Development and Local Initiative in African Savannah Environments. In Human Ecology in Savanna Environments. D. R. Harris, ed. Pp. 383–405. London: Academic.
1988   The African Experience with River Basin Development: Achievements to Date, the Role of Institutions and Strategies for the Future (Draft). Binghamton NY: Institute for Development Anthropology.
1989   River Basin Projects in Africa. Environment 31(2):4–9, 27–32.
1991   The Need and Justification for Maintaining Transboundary Flood Regimes: The Africa Case. Natural Resources Journal 31(1):75–107.

Sir Alexander Gibb and Partners; Electricité de France International; Euroconsult
1987   Study of the Management of the OMVS Common Works: Summary Reports. Dakar. OMVS.

Tabor, Joe A., and Djiby Bâ
1987   Soils and Soil Management for Agriculture, Forestry and Range in Mauritania, Mauritania Agricultural Research Project II. Tucson: University of Arizona, College of Agriculture.

Van Lavieren, B., and J. Van Wetten
1990   Profil de l'Environnement de la Vallée du Fleuve Sénégal. Arnhem, The Netherlands: Euroconsult/RIN (Institut National de Recherche pour la Conservation de la Nature).

Wane, Yaya
1969   Les Toucouleurs du Fouta Tooro (Sénégal). Dakar: IFAN.

Weigel, Jean-Yves
1982   Migration et Production Domestique des Soninké du Sénégal. (Travaux et Documents de l'ORSTOM No. 146). Paris: ORSTOM.

Woodhouse, Philip, and Ibrahima Ndiaye
1990   Structural Adjustment and Irrigated Food Farming in Africa: The "Disengagement" of the State in the Senegal River Valley. Milton Keynes, UK: Open University, DPP (Development Policy and Practice) Research Group, Technology Faculty. Working Paper No. 20.

CHAPTER 5

Shifting Social and Ecological Mosaics in Mende Forest Farming

Melissa Leach
Institute of Development Studies
University of Sussex

The Mende people who live in the rainforest environment of eastern Sierra Leone (Figure 1) create farm sites out of bush or forest and grow the culturally valued staple, rice, and a range of other food crops in a rotational bush fallow system. In this chapter I examine gender relations and changes in food production patterns over the last 30 years. Within the enduring context of the upland-inland valley swamp catena, or slope, land use has changed considerably with respect to how, and how frequently, different farm site types are used. I will argue that these agro-ecological dynamics cannot be attributed to population pressure and environmental degradation, the causal variables that outside observers commonly use to account for land use change in African agriculture. In such analyses, demographic change is seen to compel farmers to shorten their fallow periods, extend their cropping onto ecologically marginal land, and eventually to make more intensive use of less easily degraded land types. The Mende farmers I worked with are not, however, finding their land use options

135

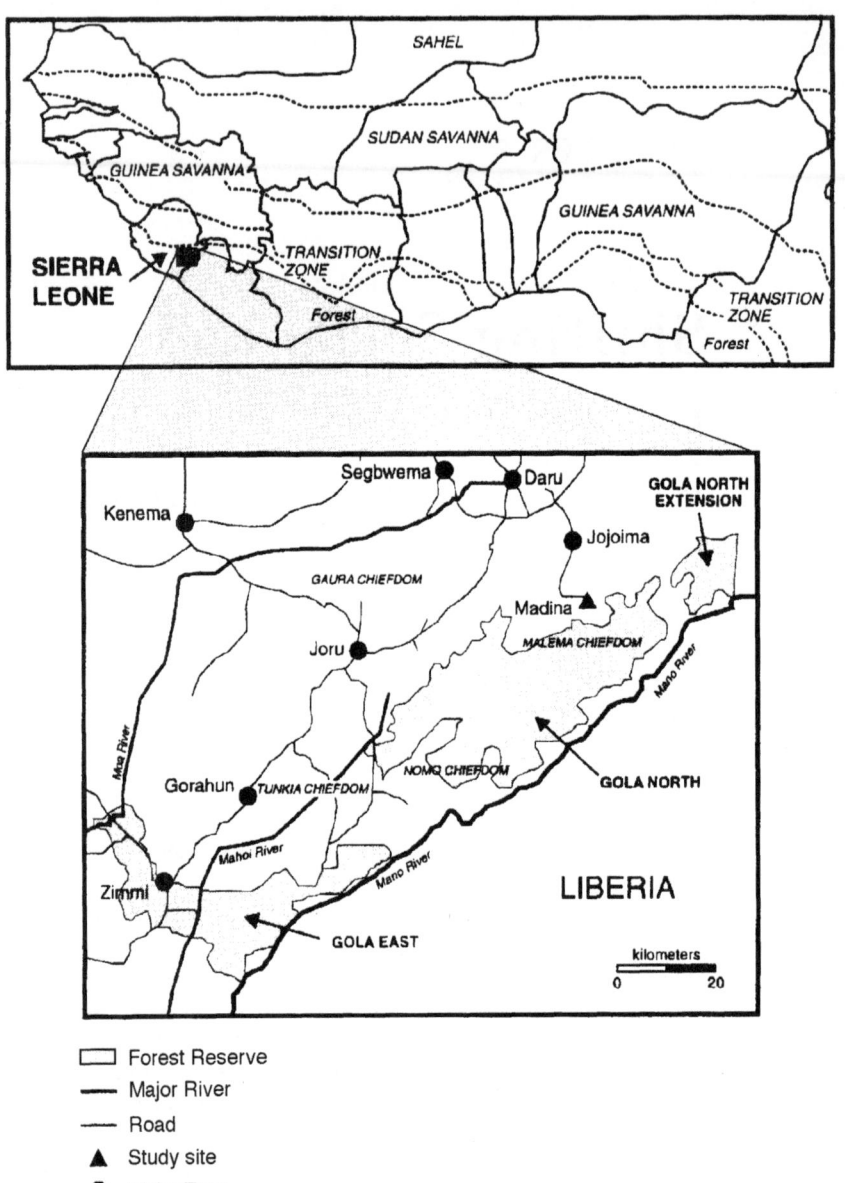

Figure 1. Location of the study site in the Gola Forest area, eastern Sierra Leone.

restricted by either population pressure or environmental degradation. Instead, the major shifts in their farming patterns are best explained by the social and gender dynamics of resource use in an evolving economic context.

The last 30 years have seen greater integration of food crop production with the larger cash economy, especially via the production of tree crops for cash. There have been changes in the gender division of labor, in land use and product control, in the terms of exchange between women's and men's activities, and in the size and internal structure of the groupings within which these activities are organized. New socio-economic conditions have altered farmers' interests and opportunities to make use of different types of farm sites, and, consequently, land use changes have occurred. In this case, I suggest that the gender dynamics of labor and resource control have been the major causal influences on the changes in farming patterns. But gender dynamics would be central to explaining land use change even in a situation in which demographic and environmental changes were placing major pressures on farmers, since they would mediate the ways that farmers responded to such pressures.

The need to study long historical processes of change in African agriculture, and the continual social negotiations that they involve, has been strongly argued by Guyer (1984a, 1986, 1988). My account of important shifts in Mende farming patterns from the pre-1960s to the present day follows this emphasis and, like Guyer's (1988) account of Beti farming change, shows a "multiplication" of labor involving changes in tasks, time, and value. There are consequences for women's and men's respective socio-economic positions as well as for agro-ecology. In some respects, the changes I will document conform to general theories of the feminization of African subsistence production. Thus, as men become more heavily involved in commercial agriculture, women's involvement in the food production sphere increases (Boserup 1970; Moore 1988). Women's concerns with independent agricultural activity increase, but their opportunities are restricted by their relative positions in changing labor and land use relations (Davison 1988; Whitehead 1991). Yet as we shall see, it is necessary to differentiate within, as well as between, the categories of "men" and "women" to see the social distributional implications of change. Although there have been important changes in Mende farming, the system also shows many enduring socio-cultural features. In this interplay of continuity and change (cf., Guyer 1984a), we must look to understand different women's and men's changing experiences of food production.

The fieldwork on which this chapter is based was carried out during 1987–88 and 1991 in Madina, a village in one of the four chiefdoms bordering the Gola forest reserves of eastern Sierra Leone. Information on modern farming patterns and concepts derives from participant observation, case studies, and surveys carried out in Madina; interviews in other settlements in the chiefdom; and selective reference to recent studies in other Mende areas. Historical reconstruction of farming during the pre-1960s era is based on oral histories and the documentary baseline provided by the anthropological studies of Little (1948a, 1948b, 1951a, 1951b, 1951c) and others during the mid-twentieth century. Interviews carried out during fieldwork in Madina fill out the changes that have occurred over the last 30 years.

FOOD PRODUCTION BEFORE THE 1960s

Bush and Farm Types in the Upland-Swamp Catena

Eastern Sierra Leone is part of a long-established zone of rice cultivation in the West African coastal forests (D'Azevedo 1962). Since at least the fifteenth century settlement and farming have converted much of the hilly, forested landscape to a shifting mosaic of forest, farm site, and bush in various stages of regrowth, offering a range of bush types with farm site potential. Mende categorize these bush types in two ways (see Figure 2): vegetationally and according to position on a catena of varying slope, soil, and moisture conditions (cf., Richards 1986). Uplands, often steep, are entirely rainfed and have free-draining gravelly soils (kOti). Lower-lying areas near the base of the catena, bului, are partially fed by runoff from upper slopes as well as by direct rainfall. Inland valley swamps below them (kpEtE) are permanently moist, and some are flooded seasonally. In vegetational terms, the Mende distinguish between forest (ngola, > 30–40 years) and various stages of bush (ndOgbO) regrowth: strong farmbush fallow (ndOgbOhinti, c. 15–30 years), fairly strong farmbush (gbOEE, c. 11–14 years), or young farmbush fallow (njOpO, <10 years). The criteria used to distinguish these stages are age of secondary succession, vegetation form, and indicator species.

Farming patterns that prevailed up to and during the mid-twentieth century were based on large intercropped catenary farms. These centered on uplands but extended down the catena to make use of the different food production sites. On the upland, local rice varieties were selected and placed to suit micro-environmental variations. Rice was

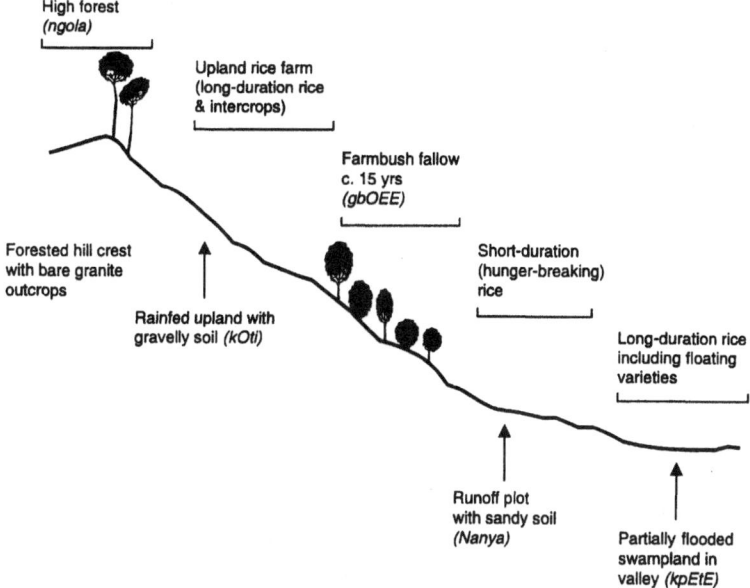

Figure 2. The upland-inland valley swamp catena. Adapted from Richards (1986).

interplanted with a wide range of other subsistence crops, including vegetables and leaves (e.g., chili pepper, eggplant, tomato), which were important as sauce ingredients, and root crops and grains (e.g., cassava, sorghum, sweet potato), which acted as seasonal hunger breakers eaten during the rainy, pre-harvest period when rice was scarce or unavailable. Wetter *bului* plots supported short-duration rice varieties, which matured in the hunger season and also acted as important hunger breakers, and a more limited range of intercrops such as maize and okra. Water-tolerant, long duration rice varieties (locally termed *yaka*) were planted where the catenary farm site extended into an inland valley swamp (*kpEtE*). Swamp soils, however, were too wet for intercrops to be grown during the rainy season.

The different stages of fallow regrowth offered particular agro-ecological advantages and disadvantages. Upland sites cleared from secondary fallow of less than 30 years could be successfully cultivated with rice only for a single year, although farmers occasionally made use of the possibility of farming upland and *bului* land for a second year with non-rice food crops (e.g., cassava, groundnuts) before leaving it fallow. Farm sites cleared from high forest gave higher yields of rice and certain intercrops (e.g., chili pepper) and could be successfully cultivated with rice a second time within five years. Clearing high forest, however, was difficult and labor demanding. Farmers normally used the more readily cleared bush of 8–15 years old, that is, the *gbOEE* or mature *njOpO* already captured within the bush-fallow cycle. Young fallow (<7 years) was normally avoided because it was insufficiently fertile and had too many weeds for rice production.

Farm Sites, Farmers and the Bush

Before the 1960s the conversion from bush to farm site, harnessing the soil fertility to grow rice, enabled settled social life[1] for the Mende. But farmers also acknowledged the fragility of the relationship between their food production operations and the forces of the bush, since productive farm sites were only a brief moment in a place's agro-ecological and social history. This fragile relationship was reflected in enduring practices and concepts that Mende farmers brought to bear on their upland farming operations. While my own fieldwork and that of Davies and Richards (1991) details these practices and concepts in the present, reference to Little's (1951a, 1951b, and 1951c) accounts suggests that they were equally present during the pre-1960s period.

Today, as well as before the 1960s, Mende conceive of farming as an annual process within which farms, as such, exist only briefly (cf., Davies and Richards 1991). Farmers enter bush (*ndOgbO*) to choose a new upland farm site in December each year. Between January and March they prepare the land by brushing (*ndoe*) undergrowth, converting the area to *ndoeke* (*ndoli yeke*; literally "new born part"). The large trees are subsequently felled (*po*) with an axe, transforming the place into *pokpaa*. The cut vegetation is left to dry and is then burned as late as possible in the dry season. If necessary, unburned branches are gathered into heaps and reburned. After burning the place is known as *motii*. As the rains set in during May through June, rice and intercrops are broadcast sown, and the soil is turned lightly with a hoe to cover the seeds. Once fully planted a site is referred to as *kpaa*, "farm." At this stage a farm hut is built for the farmers to cook and rest in, and they then await the harvest, between late October and December, with anxious expectation. Once the rice is harvested and packed safely into store for later threshing and de-husking, the farm site is referred to as *mbawoma* (behind rice). Subsequently it becomes *njOpO* (young bush) and eventually bush (*ndOgbO*) again.

Cleared farm sites are sunlit, hot, and dry, like the village, but they remain surrounded by the cold, dark bush which, although it provides the fertility for food production, also presents innumerable dangers to farmers and their operations. Along with accidents such as tree fall and climatic events such as storms and early rainfall, farmers must contend with aspects of the bush that constantly threaten to reassert themselves, thereby jeopardizing the rice and its cultivators. People engage in a constant struggle to restrain animal pests, which some-times devastate ripening crops. To do this, farmers construct stick and palm leaf fences up to a meter high with noose traps set into them at intervals in an attempt to deter cane rats, and must also keep constant guard against birds and monkeys while the rice is ripening. Weeds, which compete with the rice, must be removed. Farmers refer to all invading plants as trees (*ngulu*) regardless of whether the plant is a sapling or a herb, conceiving of weeds as reinvading aspects of the bush that will eventually comprise trees (Davies and Richards 1991).

Farmers also have a precarious relationship with the spirits of the surrounding bush. Various kinds of non-ancestral spirits (*jina*) are believed to reside in the bush, especially where there are imposing landscape features[2] or where the bush is especially dense.[3] Mende have long found that, while these spirits can bring benefits to farmers and their work, they can also cause accidents and disasters (cf., Little 1951a). In 1988 one woman told me that she feared falling asleep in her farm hut on hot afternoons lest the *tEmuisia* on the adjacent forested hill approach her with their mischievous pranks. Fallow bush and new farms made in fallow are also associated with the ancestral spirits of previous cultivators. During the mid-twentieth century sac-rifices to seek ancestral beneficence regularly punctuated the upland farming cycle (Little 1948a, 1951a). Nevertheless a farmer's status with regard to all these spirits is insecure, often likened to the position of a newly arrived stranger (*hota*) to a village. Just as a new stranger should demonstrate good intentions and allegiance to village residents in order to receive acceptance and support, farmers should "show goodness" (*fe kpEkpEya ve*) to the *jina* and ancestral spirits around a farm site if their farm work is to be successful. Thus, for example, a woman who cooks in a farm hut always calls first to the spirits of the area to share the food before dishing it out to people. Such acts seem to be motivated and shaped by people's perceptions of their insecure, negotiated relationship with bush spirits, rather than by any notion of sharing and harmonious unity between them.

Land Tenure

The ideas of farming as an annual process, and of short production cycles embedded in longer-term land histories, contributed to the concepts of land tenure that farmers have used since the mid-twentieth century. Land was not conceived of as property in the sense of something that can be owned or possessed. Instead, the two distinct concepts of landholding and land use were important.

Historically in Mende society, all the territory (*ndOlO*) of the chiefdom was held by the paramount chief (*ndOlO mahei*). Nested within this, landholding families (*mbonda ndOlO*) held areas of territory linked to particular settlements. Elderly members of the core patrilineage (*ndehu*) administered this territory on behalf of family members. Although these leaders were usually male, Mende would give control to a woman in the absence of a suitable man.[4] Landholders had power over the people and resources within the territory, having the right to use and allocate land resources and to expect political allegiance from the territory's inhabitants, but they also had obligations to protect both land and people.

Notably, landholding descent groups were also integral to local political structures. At each territorial level, descent groups were ranked according to relative socio-political influence. During the colonial period, British perceptions of each chiefdom's most important landholding family became the basis for identifying the ruling house from which paramount chiefs would be chosen (Abraham 1978).[5] At village level, key administrative and secret society leadership roles were generally occupied by members of the most powerful lineages. Local principles for ranking descent group status draw on what Murphy and Bledsoe (1986:129–130) aptly term a code of arrivals, distinguishing those high-status groups that can trace direct patrilineal descent from a settlement's firstcomer warrior or hunter-founder from those that trace descent only from a politically subordinate latecoming group, or, more weakly, through women or alternate male and female ancestors. However, the relative power of descent groups also depended heavily on achieved successes in the acquisition of clients and wealth, and was thus highly dynamic. It was not uncommon for subordinate immigrant groups to gain wealth and status and to outstrip earlier arrivals. The re-working of public versions of family and territorial histories to accommodate current political relations has long been part of the Mende socio-cultural repertoire (cf., Hill 1984; Richards 1986). One of the reasons why family and land histories are often closely guarded secrets is to avoid bringing unwanted discrepancies to light. Political relations as much as land per se are at stake in inter-family land disputes, which were arbitrated in the paramount chief's court.

At the land use level, farmers could make use of land resources (such as cultivation space, soil (*pOlO*), minerals, vegetation, water, etc.) for a temporary period. These use rights were allocated by the elders of landholding descent groups, who also arbitrated competing claims over land use. Male and female family members had the un-questioned right to be granted the use of family-held land resources for a period of cultivation. On fallowing, control reverted to the land-holding family (cf., Little 1948a, 1951a). Notably for Mende, *mbonda* (family) encompasses a much wider circle of kin and clients than the patrilineage at its core. It can include the husbands, wives, in-laws, matrilateral relatives, and clients of lineage members who all had the right to use family land, although an in-marrying wife would normally be expected to liaise through her husband, and a stranger-client through his patron. Members of other families and newly arrived immigrant strangers could also be allocated land use rights, but only in exchange for a gift (*famalo*, greeting present) acknowledging the landholders' position. The *famalo* was waived for a longer-established stranger, a signal of his incorporation as an established member of his patron's family.

Until the 1960s, farm sites were generally large, but only a small number of people engaged in land use negotiations with descent group heads each year. The extent to which different people actually took up their land use rights was shaped by the prevailing social organization of farming and by gender divisions of labor and product rights.

Organization and the Division of Labor

During the early to mid-twentieth century, groups of 40 or 50 people commonly produced rice together (Little 1948b). Between seven and ten of these groups, known as *mawEE*,[6] regularly farmed in Madina. The *mawEE* revolved around a group of brothers and sisters, their spouses, some children and grandchildren, and various more distant kin and clients, under the headship of a *mawEE-mO* (*mawEE* person).[7] As a food production group, *mawEE* members were obliged to con-tribute labor each year to make a big farm (*kpaa wa*) from which the *mawEE* head ensured their subsistence. The *mawEE* was also a resi-dential and patronage grouping, offering members many other kinds of social and economic support.[8] Male members shared large divided houses and women slept in large undivided dormitories (*pElE wa*) in village-based *mawEE* compounds, while domestic slaves were sometimes housed in more distant farming camps. *MawEE* heads were important local patrons (*numu wa*, "big person") who, as well as ensuring members' food security, offered other material assistance, e.g., paying taxes, fines, and court fees, and purchasing clothing. Cloth was given to male and female members after the completion of their respective main farming tasks (Little 1948b:43). Members' dependence on the *mawEE-mO*'s economic support assisted the *mawEE-mO*'s control over their labor. Importantly, *mawEE* heads often paid brideprice for male members, thus acquiring leverage over the labor of both the man and his new wife. Other men (including immigrant strangers) were given wives from the *mawEE* without paying brideprice but were then obliged to work for the *mawEE* to fulfill bride-service obligations (Crosby 1937). The production, patronage, and residential dimensions of *mawEE* organization were mutually reinforcing, and the annual farming cycle was a means to reaffirm them, year after year.

Farming activities and responsibilities were divided among *mawEE* members. The *mawEE* head generally secured land resource use rights for the group, and *mawEE-mO* took responsibility for decisions such as farm site location and size. A male head's senior wife (*nyaha wa*) shared (and occasionally took over) decisions about *mawEE* labor and product use (Crosby 1937). Labor was divided both by gender and by cross-cutting distinctions based on age and kinship position.

The division of labor by gender drew on strong, and historically long precedented, ideas about the suitability of certain tasks to men or women. Men created farm sites. Mende considered bush clearance tasks of brushing, tree felling, and burning to be very definitely men's work, along with building the pest control fence and farm hut. Boys learned these skills from male relatives. Task exclusiveness to men was emphasized during initiation into the men's secret society, Poro. Mende drew explicit links between bush clearance and warriorhood (Little 1951a), and the annual carving of a farm site out of the bush recalled warrior-founders' carving of territories out of the forest during settlement history. Men described clearing as "fighting the bush," often bringing their cutlasses and axes to bear with war-like aggression and cries. The intense bursts of energy and high levels of risk involved were consistent with other work thought appropriate only to men. In short, to clear bush was to emphasize one's status as a man. Clearing high forest emphasized male capability most of all as the large, buttressed trees were exceedingly difficult and dangerous to fell.

Expert tree fellers tackled huge trees with special climbing techniques and the help of magical leaf medicines (*hale*), drawing on specialized knowledge associated with (exclusively male) hunters. It was and is inappropriate for a woman to clear any kind of bush for rice production. Those who did so were labelled *hindogbahama* (a man-like woman). Generally only post-menopausal women, who were accepted as "like men" in certain respects, ever brushed fallows to make rice farms without risk of deleterious comment and accusation.

Other food production tasks were emphatically women's work, including processing and cooking the rice. Sande, the women's secret society, emphasized the female-exclusiveness of these activities. A man performing them consistently, especially in the public eye of the village, was called *nyahagbahama* (a woman-like man) and risked disapproving threats from Sande-supported women and uneasy laughter from other men.

Men initiated the annual farming cycle, whereas as rice processors and cooks, women completed it. Women cooked and brought rice to men while they worked at bush clearance and so supported them in the opening of the new cycle. Mende sometimes say of upland farming that "the bush is men's, but the farm is women's." This seems to refer both to gendered agricultural work, beginning for men in uncleared bush but for women with planting in a cleared farm site, and to the notion that men make the bush productive by converting it to a place where rice can be grown, and women make the farm productive by converting the rice to a form in which it can be eaten and used in social life.

A division of labor based on gender-specific tasks and tools combined in sequence is characteristic of many of the old world staple crops found in the West African forest zone, and such gender-labor sequences often provide means through which wider ideas about male-female relations are created and expressed (cf., Guyer 1984b). Mende emphasize that these gender-divided roles are complementary and interdependent. Neither gender's work is possible and useful without the other's. The idea that gender roles are different but balanced and complementary permeates many other spheres of Mende life, from ideas about human reproduction to the balanced powers of male and female secret societies (cf., MacCormack 1980; Bledsoe 1984). In some sense, the food production cycle provides an annually repeating locus for the affirmation of these broader ideas.

The tasks of rice sowing, weeding, pest control, and harvesting were less strongly gender-linked, and labor divisions responded more to differences of age and social status. Weeding was normally women's work, but it was not inconceivable for men to help out. High status men participated in the socially significant male tasks involved with bush clearance but would then turn to hunting and local political activity, leaving subsequent men's work such as fencing the farm site against cane rats to young men, strangers, and clients. These junior men worked with women to sow[9] and harvest the rice. There were also divisions between women. The *mawEE* head's senior wife usually organized, but took little physical part in, women's work. Female *mawEE* members sometimes divided into groups associated with each of their sleeping houses (*pElE wa*) to accomplish their tasks, each group working with an allotted complement of junior men. *PElE wa* groups oversaw the cooking of rice, in huge pots, to feed the entire group of workers (Little 1948b).

The size and labor organization of these large *mawEE* production groups was well suited to the conditions of upland catenary farming, making such a production pattern both possible and necessary. First, there was no landscape-imposed limit on the size of the upland farm site that could be cleared so the site would be large enough to feed the entire group. Second, successful upland farming required peaked inputs of group labor to fit time-bound operations to uncertain rainfalls, especially at clearing, planting, and weeding stages (cf., Richards 1986). *MawEE* labor organization made available such groups, as well as a large enough male labor force to cope with the heavy demands of high forest clearance when new land needed to be brought into the fallow cycle. If peak labor needs exceeded *mawEE* capacities, heads could draw on kinship or patron-client ties to obtain help from members of other *mawEisia* or employ a labor group. In eastern Mendeland, *kugbe*, a type of men's labor group consisting mainly of young men and stranger-clients derived from earlier warfare organization, was often used. Patrons had socio-political as well as economic interests in assembling group labor at peak periods, since this was a way publicly to demonstrate their status and to create and consolidate patron-client relations. Farm labor organization in West Africa often shows this strong political dimension (Johnny, Karimu, and Richards 1981; Linares 1981; Guyer 1984b; Berry 1989).

Interlocking with *mawEE* group rice production were two sorts of more independent food production processes. These made use of land resources nested within the *mawEE* farm site, and the products were subject to different forms of control. The first type of independent food production involved the allotting of small portions of the cleared farm site to important male, and occasionally female, dependents (Crosby 1937). This small farm (*kpaa mumu*) was generally a portion of the upland slope but occasionally a part of the swamp at the base of the catena. It was considered a separate undertaking whose labor responsibilities and proceeds rested entirely with the farmer concerned. A married man was expected to feed his wife or wives and children from his *kpaa mumu*, relying on rice from the *kpaa wa* primarily for security. Nevertheless, work on the *kpaa wa* took priority, and small farm cultivation was supposed to be carried out only in spare time left after the completion of *kpaa wa* tasks. The second independent food production type involved female *mawEE* members who planted and owned many of the intercrops. Although some hunger foods (e.g., cassava, sorghum) belonged to the whole *mawEE*, it was women who planted and owned other root crops, vegetables, and cotton, which they used for consumption and gift exchange among their kin. Women's intercropping kept them involved with each upland farm site for longer than men. For men the rice harvest signalled a conceptual boundary between food production place and regenerating bush (*njOpO*), whereas women experienced a more gradual transition. Young *njOpO* remained a source of crops, which they collected amid progressively reinvading bush plants and animals, ceasing regular visits to the place when it became too bushy and inaccessible even if the crops had not yet finished.

CHANGES IN FOOD PRODUCTION SINCE THE 1960s

Although Mende farmers were not isolated from the larger cash economy before the 1960s, their integration with it has increased in certain important respects over the last 30 years, considerably altering the context in which they produce food. In particular, the small-scale cultivation of commercial tree crops for cash has expanded. The Gola forest chiefdoms lie within the belt of Sierra Leone where cocoa, coffee, and oil palm (*Elaeis guineensis*) can all be successfully grown. Cocoa and coffee production, first introduced in the early twentieth century, became established after the Second World War and expanded under the effect of high prices during the 1960s and 1970s. Palm oil plantations added to farmers' long-established use of wild palms. The government attempted to promote improved palm varieties and techniques as a cash crop during the 1960s, and these were distributed by agricultural development agencies during the 1970s and 1980s. At the same time as these new sources of money have become available, farmers' needs for cash to purchase a wider range of goods and services has increased. These forms of intersection with the cash economy have led to renegotiations of farming organization and the gender division of labor and product control, and as a result, farmers differentiated by gender, age, and social position have new food production interests and opportunities. Farmers have responded to these new conditions in their choice of farm site type, and patterns of land use have shifted away from the near-universal intercropped upland catenary farm towards a mixture of upland farms, intensively managed swamps, and gardens.

Gender dynamics, as I will show, rather than population pressure and upland degradation, account for this intensification of swamp and garden use. This is not to imply that population size has remained static over the period in question. Comparison of the national censuses of 1963 and 1985 shows a population increase of 58% in the four chiefdoms bordering the Gola forest reserves, a relatively high rate of increase broadly comparable with other rural areas of the Eastern Province of Sierra Leone (Davies and Richards 1991). However, this increase took place from a relatively low baseline population size so that rural population densities in these chiefdoms remain about 25% lower than the regional average: about 32 persons/km^2, compared with, for example, 44 persons/km^2 in Kenema District (ibid.). Currently, farmers in the Gola forest chiefdoms do not complain of any general shortage of adequately fallowed upland farmbush, while most areas retain forest of greater than 30 years old (both within and outside forest reserves), which act as a potential buffer to any environmentally deleterious shortening of fallow periods.

Changes in Farm Household Organization

The large rice-producing groups of the 1940s are rarely found today. Most men and women produce food as members of much smaller, more numerous farm households. Normally farm households today are subunits of the *mawEE*, which is now principally a residential or patronage group. Modern small farm households may also be called *mawEE*: as functional levels have separated, so the meanings of the term, which was already polysemic, have multiplied.

Changes in farming organization are directly linked to the increasing opportunities for social and economic independence that juniors acquired through greater integration with the cash economy. As dependents acquired resources through cash crops or trade, they came to rely less heavily on the material support of their patrons and senior relatives and could therefore more easily extricate their labor from patrons' control. As one man explained, "once we made farms for a big man. But now we farm for ourselves." Little (1948a) noticed a progressive decrease in farm household size in the 1940s in Sierra Leone's Southern Province. Gola forest elders emphasize the particular role of expanding cocoa and coffee production in the 1960s. Although the first plantations were created by the heads of large *mawEE* groups, male dependents soon began to create small plantations for themselves by thinning forest or long-fallowed bush. Mende women rarely invested in tree crops on their own account for reasons explored in detail elsewhere (Leach 1990), including their limited access to male labor for tree crop maintenance and time-consuming labor obligations on their male relatives' tree crops. Cocoa and coffee incomes assisted many men to pay for their own taxes, fines, and clothing, and to support their wives and dependents. These men increasingly paid their own brideprice and could thus avoid indebting their labor so heavily to a *mawEE* head in order to marry. A man was expected to establish his own farm household at, or soon after, marriage.

There were 92 farm households in Madina in 1988 and 84 in 1991. Modern small farm households commonly consist of a husband and wife or wives, unmarried children, and sometimes an additional member such as an elderly mother. Unmarried male strangers often farm alone, and widows and separated women, once embedded in large *mawEE* farming groups, now often head their own farm households. As a result of these organizational changes, farming has become more directly linked to gender relations in marriage and to the rights and obligations associated with the conjugal contract (Whitehead 1984). Farm household headship has become integral to notions of adult male identity, to "being a man" in the sense of supporter of wives and dependents. A man who still farms with a patron or senior relative may be referred to as "only a small boy." The annual farming cycle has acquired direct associations with marriage, as a means of affirming the complementary unity of a husband's and wife's roles. Men's clearance of bush now connotes the fulfillment of married male responsibility, and failure to initiate the household farming process in this way constitutes legal grounds for divorce. Indeed for women, the phrase "he does not make farms for me" is now a common idiom to express the deterioration of one's marital relations.

The changes in farming organization have affected land use needs. Farm sites are smaller and more numerous, and there is a larger number of farm household heads who, each year, activate their rights to use land resources in order to acquire such sites. Most farm household heads can make use of a range of possibilities. Men and women living in their natal villages can use the land held by their own natal descent group or by their maternal uncles. Male and female strangers, including widows living in their deceased husband's home, can gain access to land from a spouse's or patron's lineage, while farm household heads may also beg the use of another descent group's land for the year. No farmers complain of general difficulty in gaining access to land resources, and the range of options gives most people potential access to the full range of bush and farm site types found along the upland-swamp catena (cf., Richards 1986). The shifts in land use patterns do not, therefore, simply reflect the relative access of different social groups to different land types. Family heads are sometimes called upon to arbitrate competing land use claims, however, for instance where two farmers wish to use the same piece of bush. Such cases are judged partly in terms of the relative strength of each party's claims (those of a landholding lineage member normally superseding those of a wife or stranger) and partly according to each claimant's relative political strength and social influence.

It would be wrong to see these organizational changes as a smooth transition from *mawEE* to conjugal farm household. First, farm household arrangements vary from year to year as different circumstances and dilemmas arise. For example in 1988 a recently divorced woman joined her married daughter's farm household to avoid male labor difficulties. In 1990 refugees from Liberia used a common stranger-citizen arrangement when they joined Madina farm households for the season, departing with a prearranged quantity of rice after harvest (Leach 1991a; cf., Richards 1986). As we shall see, such temporary household rearrangements are a significant source of annual flexibility in land use patterns. Second, the trend towards small farm households has not stabilized. They remain nested within the organizational possibilities of *mawEE* as production group and patrons as production group heads. Currently, powerful village patrons and *mawEE* heads often try to recruit junior kin, clients, and their wives into their farm households for the season, thus enhancing their status and authority, even when juniors would prefer to farm only with their own wives. For example in 1988 one big woman, an elder of an important landholding lineage, thus incorporated her son and his three wives so as to create a group with ten adult members, explaining that "this year we want to make a very big upland farm."

Farm Household Product Control and Independent Food Production

The (albeit unsteady) shift towards smaller, conjugal-based farm households, combined with greater integration with the cash economy, led to changes in the gender relations of food product control, consumption, and use. In the context of tensions over the use of household rice, wives and young men have become increasingly concerned with more independent forms of agricultural production and food provisioning.

As farming organization changed, household big farms (*kpaa wa*) came to be of similar size and social organization to the small farms (*kpaa mumu*) made by married male dependents in the past. The small *kpaa wa* has come to be considered the main source of food security for farm household members. The senior-most female member, usually the head's senior wife, supervises the rice store and gradually draws down the supplies as they are needed for cooking. Household rice is intended to meet members' food needs for as long as possible into the hunger season before the next harvest. But this ideal is often compromised by pressures to give rice to others, and in combination with production shortfalls this often means that supplies run out early. Rice contributions to others include political hospitality, contributions to family ceremonies, and gifts to help hungry kin and clients. The latter, made during or after harvest, have become more frequent in the modern context of many small farm households with variable annual agricultural fortunes. Although giving rice to others creates security by allowing farmers to cope by calling in their debts should their future fortunes fail, there is a balance to be struck between giving and giving too much. Husbands and wives sometimes disagree over particular gifts, considering that their partner is responding to too many of their relatives' claims. A senior wife's control over the rice store is fragile, and a farm household head can normally supersede it to remove rice to suit his own purposes.

Responsibility for filling the hunger season gap should rice supplies run out before the next harvest is supposed to lie with the farm household head. Rice can be purchased or borrowed from local traders. There are strong expectations that men should buy food with their cocoa and coffee revenues, but in practice these are often dissipated during the dry season by heavy modern cash demands (e.g., for clothing, household goods, medical expenses, and festival contributions). In these circumstances, farm household members are increasingly concerned to produce rice for themselves. Wives, unmarried sons, and male clients increasingly cultivate small, individual rice plots. As for the small farms (*kpaa mumu*) made by senior dependents under the old large *mawEE* organization, the individual concerned is wholly responsible for labor inputs and product control. Individual farmers use the rice from their separate plots to guarantee their own and their children's food security, and as independently controlled inputs into rice gift networks, which they maintain for themselves. Junior wives value independence from their senior co-wives as much as from their husband. As one said: "it is good to have your own rice if you want to cook for your own children and your friends. If you try to use the household rice your husband or co-wife might cause problems."

Women and young men have also experienced a growing need for independently produced food crops to sell for cash since, unlike most men, they lack access to tree crop revenues. Some sell rice from their individual plots during the rare periods when Sierra Leonean rice prices are favorable to rural producers. Root crops (e.g., cassava) and groundnuts have also become important sources of cash income, and women have increasingly marketed vegetables. Although root crops and vegetables for farm household consumption were and still are produced jointly by household members, personal production is considered necessary for crops that are to be sold. Women have therefore developed means to identify their personal ownership of crops planted in a shared farm site (e.g., by planting their upland intercrops in easily recognized sections, lines, and patches), and they make increasing use of separate sites.

Tree Crops, Rice and Hunger Foods

Further changes in the gender relations of food production are linked to specific dilemmas that farmers now face as a result of the competing land and labor demands of rice and tree crop production.

Tree crop cultivation has introduced new seasonal labor conflicts. These primarily involve men whose participation in the tree crop harvest coincides with their period for clearing upland farm sites, and whose tree plantation maintenance tasks (undergrowth clearance, pest control, planting new seedlings) fall at the peak of the rainy, rice production season. As one man put it, "once we worked only on rice, but our work is all scattered now." While women's obligations to help harvest and process their husbands' tree crops have increased their overall workloads, these tasks are carried out in the dry season when they hardly interfere with women's rice production work. In the modern context of small farm households, only a few patrons, with superior claims over other's labor, can invest fully in both tree crops and rice. Others resolve the rice-tree crops dilemma in various ways. Some men consistently clear farms too small to provide a sufficient store on the grounds that they can make up their food provision responsibilities to their wives and dependents by purchasing rice with tree crop revenue. But as we have seen they often fail to make such purchases, and their wives, realistic about this, privately express their wish that their husbands would concentrate on food production. Other men take a rest year from rice farming every 5–6 years to concentrate on tree farm development (cf., Engel, et al. 1984). For many farmers, each year is different as they struggle to balance dynamic organizational arrangements and obligations with fluctuating and unpredictable cocoa and coffee prices. In 1988 many men in Madina concentrated on tree crops. In 1990 and 1991 rice seemed to be back to the forefront. Much of the expansion of cocoa and coffee cultivation has taken place on low-lying *bului* land near the base of the upland-swamp catena. This has been at the expense of the short-duration rice, which was once an important hunger-breaking crop. Today, short duration rice is hardly grown.

These labor and land use conflicts between rice and tree crops are at least partly responsible for an overall decline in rice production sufficiency in eastern Sierra Leone. More specifically, it is in the context of declining farm household rice production that the increase in women's and young men's individual rice production is taking place. To compensate for rice shortfalls, some farmers show increasing interest in producing non-rice hunger foods. Some farm household heads now invest more heavily in cassava to feed their dependents, and wives produce cassava and sweet potatoes individually for themselves and their children.

Gender and Labor in Farm Household Rice Production

As well as shifts in the allocation of gendered labor between different crops and between household and independent activities, tree crop production has contributed to a renegotiation of the division of household rice production tasks between men and women.

The clearance of farm sites, linked as strongly as ever to notions of male identity, has remained definitely men's work. But rice planting and harvest, tasks once performed jointly by women and men, have devolved mainly onto women and especially junior wives. Today, a husband may claim to be fully occupied with his tree crops at the rice planting and harvesting times of year. If a husband does contribute to planting and harvesting today he is considered to be helping, not performing an expected role. It is worth noting that in the modern context of small farm households the basis for men's participation in these activities is negotiable. A man can draw on his position as a farm household head to justify leaving planting and harvesting to dependents, as the senior men in large *mawEE* groups did in the past. On the other hand, as a man may now well be the sole male member of a farm household, there is pressure to participate as junior men once did. Not surprisingly, it is older, wealthier men who seem more often to leave these tasks to their wives. Young men, stranger-clients, and farmers without tree crops seem to help more frequently. This negotiability leads to considerable variation in gender-labor patterns. Although in some farm households, in some years, the production process still involves a sequence of interdependent male and female tasks, in others it has become a process initiated by men but carried out entirely by women.

Extra-Household Labor

Changes in farm household organization and in the relationship between household and individual production have also altered farmers' needs for extra-household labor. Groups to carry out demanding operations and meet peaked labor timing needs (e.g., in upland rice production) must now be assembled from outside the farm household. Male and female farmers also need certain forms of help with their individual production activities. Yet these new labor needs intersect with significant variations in different farmers' relative access to other's labor. As we will see, these issues of labor access have a crucial influence on farmers' land use options.

Mende use a range of means to obtain extra-household labor, including drawing on kinship or patron-client ties, joining a reciprocal labor group, or paying individuals or groups by the day. Influential male or female village patrons and senior family members can call on junior kin or clients more easily than other farmers. Such patrons can also manage the social negotiations to assemble large labor groups, paying them with a combination of money, a midday meal (*kOndi*), and promised future support. Most women have more limited access to others' labor (cf., Guyer 1984b; Roberts 1988), a problem they share with young and immigrant men. Women find it difficult to assemble male labor groups, and most can draw only on a more limited range of kin ties. Women have developed arrangements to help cope with female activities. Thus, sisters, sleeping-house mates, and kitchen mates often help each other with weeding and planting, and for harvest, women link reciprocal labor arrangements with rice gift and security networks.[10] Women face severe difficulties in gaining access to male labor for tasks such as bush clearance and pest control, however. Unless she can persuade a son, a son-in-law, or lover to help, a woman must resort to paying male laborers herself for daily or task-based contracts, but willing workers are hard to find and contracts often break down.

It is especially in response to male labor access difficulties that female farmers resort to certain farm site types. Before we turn finally to these land use questions, it is useful to consider how far these socio-economic changes in farming constitute a feminization of food production and to highlight their implications for the ideology as well as the practice of gender relations.

Feminization of Food Production?

Two aspects of the changes that I have traced here suggest that Mende food production is becoming increasingly feminized. The first is the increase in women's independent production of rice and other food crops. The second is the relative increase in women's labor inputs to household food production, in comparison to men's, in terms of both tasks and time as men have concentrated their efforts on non-food commercial crops.

It would be wrong to generalize too far about this process, however. First, the changing social division of food production involvement does not follow neat male-female lines. Certain men, principally young men, immigrant strangers, and poorer farmers without tree crops, have maintained their labor involvement in food production and are, like women, now highly concerned with the production of root crops and hunger foods. Among women, it is junior co-wives, widows, and divorcees (as opposed to older and senior wives) whose labor inputs and independent food production have most increased, since such women have least access to men's resources and least opportunity to devolve their labor burdens onto others. Such cross-cutting differences of age and social status are often ignored in general theories of the feminization of food production.

Second, the changes do not take the form of universally experienced progressive trends, but of farmers' annually variable responses to changing dilemmas. For instance, studies commonly link the feminization of West African food production to the progressive break-up of family farms into more individualized production units. But the Gola forest material suggests the contextual reversibility of such processes, and it may be important to recognize the persistence of larger-scale organizational forms, which certain people can reinvoke in certain circumstances. The relationship between cash-oriented and subsistence production is not, as is sometimes implied (Klomberg and van Riessen 1983; von Braun and Webb 1987), a question of one progressively overriding the other, but a structural tension that farmers with different social obligations and resource endowments deal with in different ways from year to year. Women's labor obligations and needs for individual food crops vary from year to year in line with their shifting socio-economic and personal relationships with men and other women. This annual variability is a crucial element of Mende experience of agricultural change, yet risks being ignored when generalized social evolutionary theories are applied.

Changes and dilemmas in gendered food production relations also challenge the broader Mende ideas about gender that were annually reproduced in the upland farming cycles of the past. The concentration of women's work in an independent labor process to produce an independently controlled product, rather than a sequence of male and female tasks that results in a joint product, challenges ideas of male and female interdependence and complementarity. Women's increasing responsibility for their own food provisioning challenges notions of male identity linked to the provision of material support. And when co-wives produce food crops in separate plots, the cultural ideal of a group of co-wives working in harmonious unity in complement to their husband is called into question. Mende gender ideologies do not (yet?) seem to be undergoing major shifts in response to these challenges from the food production sphere. The challenges are buffered by broad continuities in other spheres of Mende gender relations, for example in the domain of secret society initiation and activity. But the old gender values do show signs of fragility, and people's apparent attempts to reassert them from time to time account for certain features of modern land use patterns.

Upland and Swamp Rice

The new social and economic conditions in which farmers produce food are manifested agro-ecologically in various ways. For example, land use patterns have changed as the technical possibilities linked to certain parts of the upland-swamp catena intersect with the socio-economic circumstances of particular farmers to shape their farm site choices. Moreover, where rice production is concerned, upland catenary farm sites are no longer an annual expectation. As shown in Table I, farm households in Madina now exploit a mixture of upland sites and sites using only the wettest swamp areas at the valley bottom.

Upland farming, with its peaked labor requirements, is today the prerogative of large farm households or of those with superior access to extra-household labor. Most farmers are highly concerned that their upland sites be in easily cleared fallow bush; high forest farms (*ngolagbaa*) are consequently rare today as farmers cannot easily find sufficient labor. Only two farm household heads in Madina tackled high forest in 1988. But associations between clearing this most difficult and dangerous kind of bush and male power remain as strong as ever. High forest farming demonstrates a now rare social and political capacity to assemble sufficient labor. And now that rice production and marriage are so directly entwined, some men seem to see high forest clearance – like bush clearance generally, but more so – as expressing husbandly authority and capabilities. Today such social and political issues can motivate high forest farming regardless of economic or demographic considerations. For example, one of Madina's 1988 high forest farmers felled a large area but subsequently showed more interest in hunting and village politics than in the rice. His dramatic performance appeared largely intended to demonstrate his power to the family members whose support he sought in forthcoming chieftaincy elections and to reassert his authority over his

Table I. Household Farm Site Types in Madina, 1987–1990[11]

Farm Site Type	Number of Farm Households		
	1987	1988	1990
Upland[a]	58	49	40
Swamp	29	43	35

[a]With or without extension to lower catenary *bului*/swamp areas

three wives who had intended to farm individually following recent interpersonal conflicts. The need to plant such a large area eventually brought about a concerted group effort and recreated, at least temporarily, the ideal of a group of co-wives working in tandem with their husband.

Swamp rice cultivation is becoming more common as it better suits the circumstances of certain farmers. The sequence of stages in swamp farming usually begins between April and July and ends with harvest during the following December through January. Wet, fertile swamps can be cultivated for several years in succession and using shorter fallow periods, significantly reducing the annual labor demands of vegetation clearance. Because swamp soils retain water, swamp farming operations are less time-bound. Rice transplanting deters weeds, and the piling of cut vegetation around the site during clearance operations reduces the need for pest control fencing. Swamp farming can therefore be accomplished with steadier, more evenly spread labor inputs, and thus suits small farm households and those with limited access to extra-household labor (Johnny, Karimu, and Richards 1981; Engel, et al. 1984). The gender-labor implications of swamp farming are also different from upland cultivation. Men's bush clearance in swamps is less lengthy and arduous, involving less tree felling and burning, while rice transplanting into swamps (even more than upland planting) is thought of as women's or young men's work. Male tree crop farmers tend to prefer swamp sites as these enable them more easily to reduce their labor inputs. Eleven of the thirteen farm households headed by widows and divorcees also farmed swamps rather than uplands in 1988. Although a female swamp farmer must still find male labor for the initial brushing, she need not retain it through a long sequence of clearing operations. Because she can spread out her labor inputs, she can easily intersperse her swamp work with other work demands. For similar reasons women and junior men find swamps an easier option for their additional small rice farms. Farm sites for individual rice production are almost invariably swamps.

Vegetable Gardens and Second-Year Sites

The shift from upland to swamp use for rice has direct consequences for land use options in the production of vegetables and hunger foods. When a farm household head chooses to clear a swamp rather than an upland site, female members lose their rights to plant intercrops as these cannot be grown in wet swamp soils. Women in swamp farming households have made various land use adaptations to meet their modern needs for sauce ingredients, hunger foods, and vegetables to sell for cash. First, they make arrangements to plant and harvest their intercrops on other women's upland farm sites, in effect negotiating inter-household use rights to land resources. Elderly widows and divorcees commonly plant cotton in the farms of their younger relatives. Female friends and kinswomen also make annually reciprocal arrangements to take intercrops from each other's upland farm sites. A woman will sometimes invite a close friend to join her in a major vegetable-finding expedition before a market visit, each gathering produce to sell for herself.

Second, women create separate sites to grow vegetables and hunger foods. Women have long used the areas behind their kitchens to establish a close and convenient vegetable supply. But the available space is limited and largely occupied by senior women, so new vegetable gardens must now be established in the bush. Women consider *bului* plots ideal, but as we have seen, cocoa and coffee cultivators have occupied most of these. Thus constrained, women create gardens on two types of rice production sites, introducing vegetables as a separate stage in the land use-fallow sequence. They make the first type of garden in swamp land only during the dry season, abandoning the site to fallow or swamp rice production in the rains. Women create the second type during the rainy season by recultivating a recently rice-farmed upland site (i.e., *njOpO* of 1 to 2 years old). Occasionally women cultivate both complementary types in the same year, transferring their crops and their attention from the upland to the swamp in the dry season (Leach 1991b). Women find separate vegetable gardens well suited to market production as they provide for greater and more predictable production and clearer personal rights over produce than when vegetables are planted as intercrops (cf., Klomberg and van Riessen 1983).

Farmers have also intensified their use of upland farm sites for a second or third year before they are left fallow. Women and young men increasingly produce groundnuts on such second-year sites, while farm household heads and individual women use them to plant cassava and other root crops as hunger foods and for cash sale, especially if they lack access to an upland intercrop site. Because cassava and other root crops have low and flexible labor requirements, an individual can, if necessary, manage them alone. And because indigenous varieties can thrive on relatively infertile second- and third-year soils, they obviate the need to carve an entirely new site from bush or forest.

This is an advantage for women and for farmers with limited access to male labor. Furthermore, unlike bush clearance for rice farming, second- and third-year site preparation is not a socially loaded key male task. Women can acceptably brush away the low vegetation regrowth in the *njOpO* for themselves. Pest control remains a problem, and some women do not embark on such projects if they cannot find a man to fence their cassava or groundnut plot against cane rats (*Thryonomys swinderianus*). But others manage to make mutually beneficial arrangements with male kin and friends. For example, two women and three young men grouped together to perform the various male and female tasks in preparing, protecting, and planting a large cassava and groundnut site in 1988, subsequently designating, tending, and harvesting their own plots within it.

Socio-Economic and Ecological Implications of Land Use Changes

Although the causes of these land use changes are primarily socio-economic rather than demographic or environmental, they have certain negative implications that are likely to become more serious as population densities rise. The changes give scope for new kinds of conflict over land resource use that may well intensify in the future as growing populations place increasing overall pressure on land resources. Throughout eastern Sierra Leone there is growing competition among swampland uses and users for rice production, vegetable gardening sites, and tree crop development. Similar pressures may soon affect upland use for different types of extended cropping, such as for varied annual crops on second- and third-year land, and for tree crop development. There is as yet no evidence of overall strain on the processual framework within which Mende consider land tenure (cf., Davies and Richards 1991), but people with weaker claims or lesser political influence may well lose out in the competition for access to increasingly scarce and valuable sites. Recent cases in which strangers' swamp use has been revoked in favor of lineage members testify to this. Women and junior men may also find their in-nested, secondary land resource use rights, such as to cultivate intercrops and gardens in other's farm sites and land use-fallow sequences, increasingly undermined.

These patterns of site-use and cropping also have ecological implications. They imply shifting pressures on soils and vegetation within different parts of the upland-swamp catena, over and above the general changes in soil and vegetation composition that are thought to occur with successive cultivation cycles (Davies 1990). Ecological changes resulting from farming are felt more strongly where population densities are higher and fallow periods generally shortened, as in parts of central Sierra Leone (Davies and Richards 1991). Even in the Gola forest area the growing pressure on certain types of bush and farm site is leading to non-cyclical ecological change. This is true of inland valley swampland, whereas in the Guinea savannas of northern Sierra Leone, Nyerges (1989) has suggested that second- and third-year plot cultivation damages the potential for bush fallows to recover through coppicing. Nyerges' findings concerning vegetation dynamics may well have some validity in the Gola forest area, although they would need to be modified to account for the particular cultivation and tree management techniques as well as the more humid conditions prevailing in these southerly bush fallow systems.

CONCLUSIONS

My concern with ecology in this paper has not involved analysis of precise soil and vegetation dynamics and their consequences, either in terms of measurements that outside scientists might use or criteria that the Mende might employ. Rather, I have looked at how, and how frequently, different bush and farm site types are used. I have shown that environmental management and changes, including those that ecologists might see as having negative or deleterious implications for the sustainability of agro-ecosystems, need to be understood in terms of different resource managers' interests and opportunities and the dynamic relations between them. Even regardless of the complex nature of ecological dynamics themselves, environmental pressures are mediated through a complex of relations and factors influencing site choice and cultivation patterns that ensure that they will be unevenly manifested in space, time, and across different social groups. It is difficult to capture this variability and multidirectionality unless food production is examined in relation to people's differentiated experiences on the one hand, and over the long term on the other. This has involved a disaggregated approach to the economy within which resources of various kinds are managed, distributed, and used, seeing beyond, within, and across categories such as men and women, or particular social units (e.g., households), to the changing configurations in which people combine themselves and resources in different circumstances.

The activities, exchanges, combinations and recombinations, and dilemmas people engage in when managing resources go far beyond the production of food. They also reflect and reproduce wider cultural ideas and social and political relations, whether these concern the relationship between farmers and the bush or the proper organization of marriage. As Guyer (1984b) argues, agricultural production cycles not only describe the material means of subsistence, they are also a symbolic means of validating social arrangements. From day to day, resource use shapes and is shaped by interpersonal relations and by the ways people get on with and think about each other. As new economic and ecological dilemmas are thrown up by processes of change, provoking creative responses, so new interpersonal dilemmas emerge. In this Mende case I have highlighted, for instance, the ambiguous relationship between more independent forms of food production and relations of authority at *mawEE* and conjugal levels, on the one hand, and the ways that women's and men's current range of separate production involvements create new strains in the conjugal contract, on the other (cf., Whitehead 1984). Recent studies of food production in West Africa emphasize how changes in production conditions (including those induced by the state) can bring experiences of material life into contradiction with dominant cultural representations, thereby precipitating struggles over meaning (Berry 1989; Carney and Watts 1990). In the Mende case such struggles are evident but they are accommodated within a broad and flexible set of concepts of people-bush relations, gender interdependence and complementarity, and land tenure relations, which show no signs of fundamental change. The dominant sense is of subtle shifts within an enduring repertoire (cf., Guyer 1988).

Glossary

This glossary contains all Mende words cited in the text more than once. The phonetic symbols ɔ , ɛ and ŋ in Mende are designated using the capital letters O, E, and N.

bului	Low-lying, runoff plot at foot of soil catena
gbOEE	Strong fallow farmbush, c. 11–14 years old
jina	Non-ancestral/nature spirit
kpaa	Rice farm
kpaa wa	Big/household rice farm
kpaa mumu	Small/individual rice farm
kpEtE	Inland valley swamp, swamp rice farm

mawEE	Past: residential/farming group. Present: i) group of people who seek protection of same patron in town business; ii) residential household; iii) farm household
mawEE-mO	*MawEE* head
mbonda	Family: consanguines, affines, and residential affiliates
ndehu	Lineage
ndOgbO	Bush
ndOlO	Land, country, territory, world
ngola	High forest, > 30–40 years old
nyaha	Woman, wife
pElE wa	Women's sleeping house

ACKNOWLEDGMENTS

This chapter is based on fieldwork carried out in eastern Sierra Leone during 1987–88 and 1991. My grateful thanks are due to the people of Malema Chiefdom, especially Madina; to my research assistant and colleagues; and to my funders (the Economic and Social Research Council and the MacArthur Foundation via the Institute of Development Studies "Securities" project). Responsibility for the opinions expressed here of course remains mine alone.

END NOTES

1. Rice is essential for social life; not only is it central to cultural notions of dietary adequacy, considered as the only real food (*mEhE vuli*), it is also integral to processes which bind and legitimate social relations, and can in this sense be envisaged as a kind of social glue (Linares 1985). Rice is shared to express friendship and love, and is central to political hospitality, funerals, and initiations.

2. For example *tingOi* and *njalOi* are found in pools and waterfalls, and dwarf-like *tEmu* on large hills and rocks.

3. For example *ndOgbOjusu* inhabits the deep bush (*susu/jusu* = deep recess).

4. In such circumstances the woman was considered merely a temporary custodian, and land control was expected to revert to male hands as soon as possible. In practice, lineage segments and landholding sometimes remain under female leadership for several generations.

5. In some, but by no means all, chiefdoms, ruling houses and their incumbents are colonially derived impositions with little basis in local legitimacy.

6. From *mu wElE*, literally "our house" (Little 1948b).

7. Although most *mawEE* heads were male, influential senior women could acquire the title by inheriting it from a dead husband or male relative in the absence of suitable male heirs (Little 1948b:40).

8. The term *mawEE* seems to have originated during the turbulent warfare times of early forest settlement. It refers to the entire group of male and female dependents and slaves whom a warrior protected and accommodated (Little 1948a, 1948b; Abraham 1978). The main social features were retained after the decline of warfare and slavery at the end of the nineteenth century.

9. Female participation in rice sowing and hoeing-in in this area of Mendeland probably reflects socio-cultural influences from nearby Liberia, where planting was and is primarily a female role. In Mende areas to the northwest, it was men who sowed and hoed-in the rice, and women merely followed behind to clear up stray plant debris (Little 1951b).

10. A woman invites her female kin and friends to cut the rice she has helped plant, paying each helper 3 to 4 "ties" of rice per day, expecting to be invited back to the helper's farm when rice is harvested there. A tie typically provides for 1 to 2 family meals.

11. I have no data for 1989. The varied total numbers of farm households each year reflect the year-to-year organizational rearrangements discussed earlier, as well as in- and out-migration.

REFERENCES

Abraham, Arthur
    1978   Mende Government and Politics under Colonial Rule: A Historical Study of Political Change in Sierra Leone 1890–1937. Freetown: Sierra Leone University.

Berry, Sara
    1989   Social Institutions and Access to Resources. Africa 59(1):41–55.

Bledsoe, Caroline
    1984   The Political Use of Sande Ideology and Symbolism. American Ethnologist 11:455–472.

Boserup, Ester
    1970   Women's Role in Economic Development. London: George Allen and Unwin.

Braun, J. von, and P. Webb
    1987   Effects of New Agricultural Technology and Commercialization on Women Farmers in a West African Setting. Washington: International Food Policy Research Institute (IFPRI).

Carney, Judith, and Michael Watts
    1990   Manufacturing Dissent: Work, Gender and the Politics of Meaning in a Peasant Society. Africa 60(2):207–241.

Crosby, K. H.
  1937 Polygamy in Mende Country. Africa 10(3):249–264.

D'Azevedo, Warren L.
  1962 Some Historical Problems in the Delineation of a Central West Atlantic Region. Annals of the New York Academy of Sciences 96:512–38.

Davison, Jean, ed.
  1988 Agriculture, Women and Land: The African Experience. Boulder and London: Westview.

Davies, Glyn
  1990 Survey of Agro-Forestry Potential of Indigenous Trees in Sierra Leone. Consultancy Report to Bo-Pujehun Rural Development Project.

Davies, Glyn, and Paul Richards
  1991 Rainforest in Mende Life: Resources and Subsistence Strategies in Rural Communities around the Gola North Forest Reserve (Sierra Leone). Report to ESCOR, UK Overseas Development Administration.

Engel, A., et al.
  1984 Promoting Smallholder Cropping Systems in Sierra Leone: An Assessment of Traditional Cropping Systems and Recommendations for the Bo-Pujehun Rural Development Project. Unpublished paper, Technical University, Berlin.

Guyer, Jane I.
  1984a Family and Farm in Southern Cameroon. Boston: Boston University African Studies Center.
  1984b Naturalism in Models of African Production. Man (N.S.) 19:371–388.
  1986 Intra-Household Processes and Farming Systems Research: Perspectives from Anthropology. In Understanding Africa's Rural Households and Farming Systems. Joyce Lewinger Moock, ed. Pp. 92–104. Boulder and London: Westview.
  1988 The Multiplication of Labor: Historical Methods in the Study of Gender and Agricultural Change in Modern Africa. Current Anthropology 29(2):247–259.

Hill, M.
  1984 Where to Begin? The Place of the Hunter-Founders in Mende Histories. Anthropos 79:653–656.

Johnny, Michael, John Karimu, and Paul Richards
  1981 Upland and Swamp Rice Farming Systems in Sierra Leone: The Social Context of Technological Change. Africa 51:596–620.

Klomberg, Alphons, and Agatha van Riessen
   1983   Marginalization of Export Crop Producing Households, Exempli-
          fied by the Position of Women, Food Crop Production and Market-
          ing Conditions: Upper Bambara Chiefdom, Sierra Leone, As a Case
          Study. Unpublished M. A. Thesis. Nijmegen.

Leach, Melissa
   1990   Images of Propriety: The Reciprocal Constitution of Gender and
          Resource Use in the Life of a Sierra Leonean Forest Village. Ph.D.
          Thesis, University of London.
   1991a Dealing with Displacement: Refugee-Host Relations, Food and For-
          est Resources in Mende Communities of Sierra Leone During the
          Liberian Influx, 1990–91. IDS Research Report 22, Brighton, Sussex.
   1991b Social Organization and Agricultural Innovation: Women's Vege-
          table Production in Eastern Sierra Leone. *In* Proceedings of a Wor-
          kshop on Peasant Household Systems: Partners in the Process of
          Development. Hartwig de Haen, ed. Pp. 186–208. Deutshe Stiftung
          für Internationale Entwicklung (DSE): Germany.

Linares, Olga F.
   1981   From Tidal Swamp to Inland Valley: On the Social Organization of
          Wet Rice Cultivation among the Diola of Senegal. Africa
          51:557–95.
   1985   Cash Crops and Gender Constructs: The Jola of Senegal. Ethnology
          24:83–94.

Little, Kenneth L.
   1948a Land and Labour Among the Mende. African Affairs 47(186):23–30.
   1948b The Mende Farming Household. Sociological Review 40:37–55.
   1951a The Mende of Sierra Leone: A West African People in Transition.
          London: Routledge & Kegan Paul.
   1951b The Mende Rice Farm and its Cost, Part 1. Zaire 5(4):227–73.
   1951c The Mende Rice Farm and its Cost, Part 2. Zaire 5(4):371–80.

MacCormack, Carol
   1980   Proto-Social to Adult: A Sherbro Transformation. *In* Nature, Cul-
          ture and Gender. Carol MacCormack and Marilyn Strathern, eds.
          Pp. 95–118. Cambridge: Cambridge University.

Moore, Henrietta L.
   1988   Feminism and Anthropology. Cambridge: Polity.

Murphy, William P., and Caroline Bledsoe
   1986   Kinship and Territory in the History of a Kpelle Chiefdom (Liberia).
          In The African Frontier: The Reproduction of Traditional African
          Societies. Igor Kopytoff, ed. Pp. 123–147. Bloomington: Indiana
          University.

Nyerges, A. Endre
  1989   Coppice Swidden Fallows in Tropical Deciduous Forest: Biological,
          Technological, and Socio-Cultural Determinants of Secondary For-
          est Successions. Human Ecology 17:379–400.

Richards, Paul
  1986   Coping with Hunger: Hazard and Experiment in an African Rice
          Farming System. London: Allen and Unwin.

Roberts, Penelope A.
  1988   Rural Women's Access to Labour in West Africa. *In* Patriarchy and
          Class: African Women in the Home and the Workforce. Sharon
          B. Stichter and Jane L. Parpart, eds. Pp. 97–114. Boulder and Lon-
          don: Westview.

Whitehead, Ann
  1984   I'm Hungry, Mum: The Politics of Domestic Budgeting. *In* Of Mar-
          riage and the Market: Women's Subordination Internationally and
          Its Lessons. 2nd ed. Kate Young, Carol Wolkowitz, and Rosylyn
          McCullagh, eds. Pp. 93–116. London: CSE Books.
  1991   Food Crisis and Gender Conflict in the African Countryside. *In* The
          Food Question: Profits Versus People? Henry Bernstein, et al. eds.
          London: Earthscan.

CHAPTER 6

The Social Life of Swiddens: Juniors, Elders and the Ecology of Susu Upland Rice Farms

A. Endre Nyerges
Anthropology/Sociology Program
Centre College, Danville, Kentucky

This chapter concerns the individual patterning of farmer success and failure among Susu agriculturalists in northwestern Sierra Leone. In it, I analyze production problems in their social and environmental contexts and adopt an approach that I term the "ecology of practice" (Nyerges 1992). This approach to the study of production and re-source exploitation focuses on how local environmental problems, such as food insufficiency and resource decline, are produced through social action. The ecology of practice emphasizes the incorporation of resources into the social interactions of farmers and, in particular, examines the environmental and agricultural consequences of indivi-dual efforts to achieve and maintain socially defined management goals in the context of established systems of hierarchy. In developing this approach, my concern is ultimately to clarify the relationship between social practice or action, on the one hand, and processes of ecological adaptation or coping, on the other. For example, in the case of swidden agriculture we know from decades of research that the local environmental knowledge of swiddeners is considerable, that the technology of swiddening is well suited to local environments, and that swiddening is highly efficient and conservative of resources (Conklin 1957; Geertz 1963; Richards 1986). In a word, it is adaptive, or to follow a terminological tendency evident in much recent literature, it can constitute a basic coping strategy of a local population. Yet the application of this technology in specific sites by farmers in any system of resource management may be modified by many additional factors, including social and organizational ones, in particular.

For ecological anthropology, a theory of resource management and exploitation is needed that incorporates factors of both coping and practice. Students of pastoralism, for example, have distinguished between herder and husbandman; between the mechanics and techniques of flock logistics and day-to-day herding labor, on the one hand, and the organizational factors of long-term strategy, access to resources, and accumulation of surplus capital, on the other (Paine 1972; Nyerges 1982). Thus, a distinction exists in the literature between the technology of management, which may be ecologically adaptive, and the organization of management, which, whether adaptive or not, is oriented to and constrained by social goals and hierarchies. This distinction accords with Spooner's (1982:401–402) definition of approaches in ecology as largely either "technocentric," focusing on technology and its use and impact on society and the environment, or "sociocentric," focusing on conflict and power relationships among groups and individuals of different statuses in society.

In a series of recent publications (Richards 1983; Vayda 1983; Huss-Ashmore 1989), microlevel approaches to the analysis of resource management systems and hazard avoidance are advocated for the study of small-scale agricultural economies. Richards (1983, 1986), for example, advocates an approach characterized as "environmental particularism" in the study of African land-use systems, in an effort to properly appreciate "the abundant evidence of small-scale food producer's interest in and commitment to technological change" (1986:2). The historical achievements of African agriculture include zero or minimal tillage, selection of drought- and pest-resistant crops, use of ecologically complementary plant combinations, multiple planting of single cultivars, planting of multiple crop varieties, and intercropping (Huss-Ashmore 1989:27–28). Richards' perspective is that if researchers attend to local farming practices they can expect to discover new technological adaptations, such as particular multi-crop combinations, which increase productivity under particular circumstances. The overall argument is that local adaptations reflect long-term trial and error experimentation and innovation, based on sophisticated environmental knowledge and management strategies. As Richards states, his goal is to provide the option "of building on the best of local initiatives and supporting changes already taking place within the peasant farming community" (1986:2). Thus, in carrying out programs of agricultural development in local areas, planners need to enhance the system in place rather than import replacements.

Although I agree with the point of view expressed, I also find that local production systems – such as those I am familiar with in northwest Sierra Leone – are characterized by social asymmetries that are strongly institutionalized and that in many cases lead to management variations and failures (Nyerges 1987, 1992). The ecology of practice reflects this concern with the sociocultural dimensions of problems of resource exploitation, that is, with variations in the actual ability of farmers to meet the organizational challenges of carrying out successful farming adaptations in their fields. In other terms, swidden farms, as much as the farmers who make them, may be said to lead social lives (see Appadurai 1986; Kopytoff 1986). As resources, they are incorporated into the practices of individual farmers who are engaged in a social competition. Therefore, understanding the ecology of these swiddens (as any resource) requires understanding them in the context of the social lives of the farmers who make them and of the social systems in which these farmers live and operate. As I have noted elsewhere in defining the ecology of practice (Nyerges 1992:873), "ecological anthropologists interested in issues of resource management will find most useful those hypotheses that allow them to relate people's positions in local hierarchies to the exploitation of the natural resources on which they depend."

The Susu of Sierra Leone are small-scale agriculturalists characterized by low population density and high dispersion, a high degree of internal social differentiation, and a corresponding ideology of patriarchal and gerontocratic control. In previous publications, I have examined Susu swiddening from the perspectives of the seasonality of the Guinea savanna environment, the ecology of secondary successions in tropical deciduous forest, and the culture history of frontier migration in the western African region (Nyerges 1988, 1989, 1992). I have further discussed the impending ecological changes expected to be brought about in this region by the disease-clearing efforts of the World Health Organization's Onchocerciasis Control Programme (Nyerges 1987). In the present chapter, however, I examine Susu swiddens primarily in terms of their cultural biographies, that is, from the perspective of the social lives of the swiddeners who produced them. In the particular case of the Susu of Kilimi, social categories that differentiate among persons include men and women, elders and juniors, "citizens" and "slaves" (or the descendants of slaves), and landowners and newcomers. Each of these categories of actors, given a system of differential access to and control over the means of production, might reasonably be expected to manage resources to some degree differently from one another, to be more or less successful as farmers, and to produce swidden farm sites that vary correspondingly in yields and in ecology.

In what follows, I seek to deal with the related dimensions of coping and management in Susu farming. I first consider Susu population dynamics and the Susu relationship to the environment in terms of land availability, in order to demonstrate that arable land per se is not in short supply and not directly a cause of farming failure or environmental difficulty. Then, I discuss basic social arrangements characteristic of Susu village life and farming activity, in order to show that seasonal labor availability varies among farmers and overall is in short supply. I contrast Susu ecology to the coping mechanisms and social stratagems of Mende farmers elsewhere in Sierra Leone in order to show that, for example, Susu yields are variable, seasonal deprivation is prevalent, and that Susu farm site choices are based as much on social and organizational factors as on agro-ecological ones. Finally, I analyze the productivity of Susu farms in relation to the social positions of the farmers who produced them, in order to examine the related dimensions of coping, management, and social hierarchy in Susu swidden ecology.

POPULATION DYNAMICS AND THE AVAILABILITY OF ARABLE LAND

The Susu of Kilimi (Figure 1), with whom I lived and worked for 17 months between 1981 and 1984, number 535 to 600 persons living in an area of 240 to 288 km^2, for a population density of 1.9 to 2.5

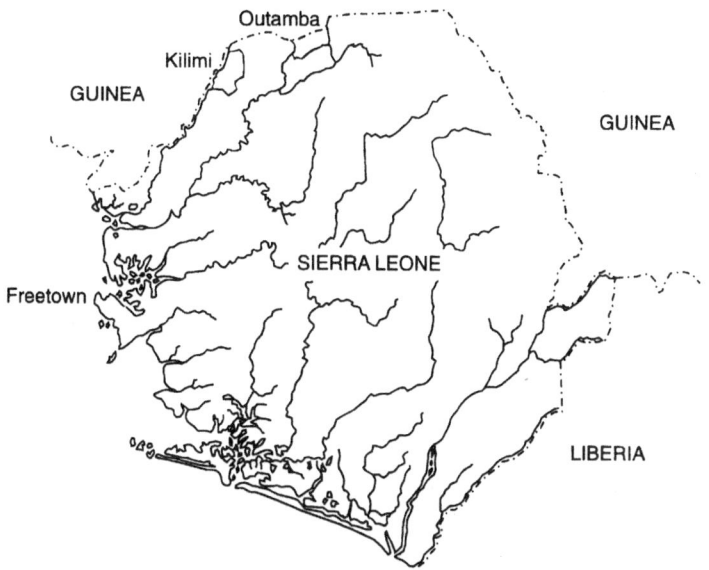

Figure 1. The location of the Kilimi study area. The proposed Outamba-Kilimi National Park, Sierra Leone.

persons per km². The density for Tambaxa Chiefdom, where the study area of Kilimi lies, is approximately 10 persons/km². By contrast, rural population densities for Sierra Leone as a whole are approximately 40–50 persons/km².[1] This extreme low density is maintained in part by factors of malnutrition and disease, which in turn may affect fertility. Thus, few Kilimi women have many surviving offspring, and a significant percentage are childless. Sexually transmitted diseases are prevalent in the region, and parasitic diseases, such as malaria, onchocerciasis (river blindness), and filariasis (elephantiasis), are also common. Medical resources are few, and local herbalists, although effective, are unable to treat many problems. At the same time, emigration is not a factor, although at the time of research it was common throughout much of the rest of Sierra Leone as rural young men left their homes to work in the cities or the diamond mining areas. In contrast, Kilimi young men, while more mobile than other Kilimi residents, typically traveled out of the area for strictly short-term or seasonal purposes, returning to the village to farm, to inherit, or to marry (for a more detailed discussion of Susu demographic factors, see Nyerges 1992:863–865).

This Kilimi Susu population occupies a region of transitional Guinea savanna/moist deciduous forest mosaic. Farmers exploit the environment for the swidden farming of upland rice and intercrops, employing techniques of minimal cultivation and a technology of locally crafted cutlasses (machetes), axes, and hoes to clear and farm the land. According to preliminary aerial photo analysis covering approximately 209 km^2 of park land, Kilimi plant associations include an estimated 74% savanna woodland, 14% lateritic hardpan, 8% deciduous or riverine forest, and 4% boli (seasonally flooded grassland). Of these areas, only lateritic hardpan, which is virtually devoid of soil, can be considered to be non-arable, thereby slightly decreasing the available Kilimi land area and marginally changing the effective population to land ratio.

Preferred farming land of moist forest fallows covers only 8% of the Kilimi area, for a spurious population density of approximately 32 to 35 persons per km^2 preferred land. Yet bolis are usable given available techniques and tools. Furthermore, the category of savanna woodland includes a substantial, albeit unquantified, portion of land that is better described as mixed savanna woodland and deciduous forest, including trees typical of both savanna and moist forest areas. This formation has a closed canopy, limited grass understory, and good soil development suitable for upland rice and other swidden crops. Much of the rest of the savanna woodland category is also ecologically suited for the production of groundnuts or fields of sorghum or millet. Farmers prefer the most favored crop, upland rice, and the easiest land to use, 30-year-old fallow in deciduous or riverine forest and the small areas of inland valley swamp within these sites. These preferences, however, do not alter the fact that a larger population could increase production by shortening fallows, expanding the area under cultivation, and switching to other staples (sorghum, millet), even if the inevitable result would be lower returns on labor and accelerated environmental decline. Thus, effective land availability must be understood as a function of technology, management, and cultural preference. Several options are available to Susu farmers, and land is not in short supply.

THE SEASONAL SUPPLY OF LABOR

Instead, the crucial problem in Susu farming is the differential availability of labor to various farmers, particularly during the agriculturally productive rainy season. In the Kilimi environment, rainfall is sharply differentiated between a virtually rainless dry season from November to March, followed by heavy, often extreme rainfall from April to October. Growth of plants occurs only during the five- to seven-month period of the rains, and all agriculture in the region is geared to this cycle (Nyerges 1988: Table 1). The relationship is rigid, and for the farmer, timing is all-important. The swiddens must be felled and allowed to dry sufficiently before burning at the last possible moment before the rains begin, and rice must be planted in good time to mature before the rains end. Under the harsh conditions of the Guinea savanna environment, the hunger period of the rains before the annual harvest is felt by even the most successful farmer, and failure to organize the labor necessary to prepare fields in a timely and efficient manner means loss of crops. This need to meet a seasonal schedule for production, when the water balance is positive and plant growth can occur, entails many significant opportunity costs for farmers and poses extreme time and labor constraints on productivity. Common adjustments made by farmers are the division of labor by gender and the seasonal formation of young men's cooperative work groups. By these means, household labor is organized to carry out the tasks of weeding, guarding against bird and mammal crop raiders, harvesting, and threshing, and young men are mobilized to perform the strenuous late dry season to early rainy season tasks of felling, clearing, and planting swiddens.

Along with technological adaptations of minimal tillage, use of multiple crop varieties, and intercropping, these labor mobilization strategies enable Susu farmers to cope with the seasonal environment sufficiently well to produce good yields, on average, from their farmland. However, labor adjustments of both household and work group organization are subject to manipulation in Susu society. Consequently, labor is not equally available to all farmers, and the Susu in fact experience numerous problems in maintaining the resource base and meeting subsistence needs. In particular, close analysis of yield data from Susu swiddens indicates a substantial and patterned variation among individual farmers in the actual ability to produce a successful crop and to store adequate yields to stave off hunger and acquire cash.

THE DIMENSIONS OF THE PROBLEM

Although aggregate yield statistics on the Kilimi agricultural system conceal the important farmer-to-farmer variations that are the main subject of this chapter, mean yield figures, nevertheless, are useful in comparing Kilimi with other sites and in establishing the dimensions of the problem.[2] Thus, for the 1983 study year, my 31 samples of upland rice harvests indicate a mean yield of 1225 kg/ha and a mean yield-to-seed ratio of 19:1, while five samples of swamp rice indicate a mean yield of 3680 kg/ha and a mean yield-to-seed ratio of 98:1. These returns are apparently equivalent to Mende upland and swamp rice farm yields estimated by Richards (1986:104) for the same farm year (1983) elsewhere in Sierra Leone. For Mende rice farms in the rainforest region of central Sierra Leone, Richards estimates yields of roughly 1000 kg/ha (and a yield-to-seed ratio of 17–20:1) for upland farms and 2500 to 3000 kg/ha (and a yield-to-seed ratio of 40–50:1) for swamp farms.

This apparent comparability suggests that, at least in terms of average yields per hectare, Susu harvests are as good as, or perhaps even slightly better than, harvests in other places. Yet this comparison does not in itself show whether total yields are sufficient for subsistence, nor does it indicate the extent to which farmer-to-farmer variability occurs in either system. Indeed, as presented the Susu-Mende comparison is misleading, unless we make a methodological distinction concerning how the yields were obtained. Thus, Richards' estimates of Mende rice yields are based on *farmers' post-harvest reports* of stored rice. In other words, the Mende data reported in *Coping with Hunger* reflect total yield from the farm already reduced by any harvest-time consumption and redistribution. By contrast, the Susu yield data are based on *direct sample yields* taken from fields before any harvest by the farmer or redistribution to others. For the Susu of the study village, the data for *reported post-harvest stored yields*, which are the figures I believe to be most comparable to Richards' Mende yield estimates, are mere fractions of the total yields that I estimated on the basis of sample harvests and far short of the Mende reports of stored harvest yields. Figure 2 compares estimated total rice yields based on my sample harvests to reported stored harvests for the Kilimi study villagers. As it shows, the portion of crop reported stored ranged from 0% to less than 45%, averaging 15%, of estimated total yield. Overall, the Susu stored yields from upland rice farms averaged a mere 200 kg/ha, or 20% of the comparable Mende figure for yield at harvest time (i.e., 1000 kg/ha).

The magnitude of this difference between Susu and Mende stored harvests is obviously great and bears some further explanation. According to Richards (1986:104), the Mende store most of what they harvest, cope successfully during the preharvest hungry season, and

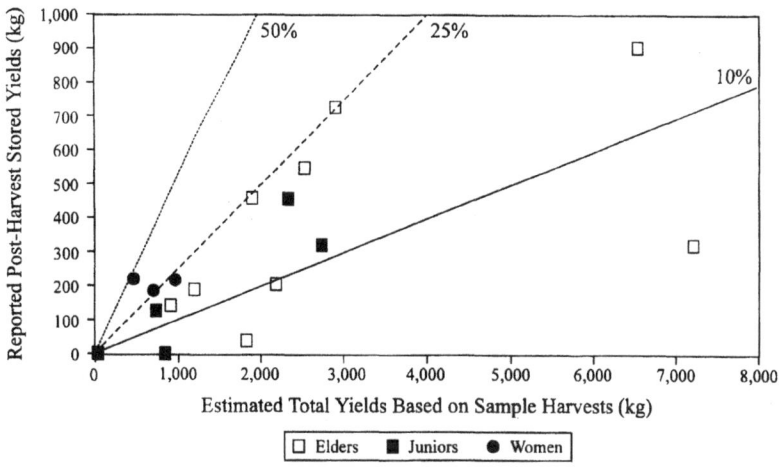

Figure 2. Portion of total yields reported stored at harvest by study village farmers.

also have about 25% of their harvested yield available to sell. As he states, "The quiet early evening negotiation for the purchase of a 'surplus' pan or two of rice from the wife in charge of food preparation for the day is a regular feature of the life of the Mogbuama produce trader" (1986:66–67). He further attributes the occurrence of occasional individual farming failures to being too hard up to mobilize sufficient labor to farm properly and to various mistakes, errors of judgment, or "lessons of experience" (1986:111–112).

The Susu, however, consume or redistribute virtually all of their crop directly off the field during the September through December harvest period. They have no rice to sell, and remaining stored rice and other grains are largely required to feed work groups. Although their swidden farming system incorporates elements of minimal tillage, exploits multiple varieties of cultivars and adaptive polycropping, and provides farmers with a variety of site choices, farmers nevertheless face both a substantial deficit in food production and a significant hungry season during the rains before harvest. Thus, during the 1983 rainy season I was approached by a delegation of Kilimi study village elders asking me to send to Freetown for two large sacks of rice to sell to the villagers at cost. When the 72 kg (158 lbs) of Texas rice finally arrived, via headload from the chiefdom capital as no Kilimi roads are open in the rains, the full quantity of rice was purchased, cup by cup, within days. I was later assured that, while no one would have starved, without this end-of-the-rains food aid the people would have been seriously weakened and very hungry indeed.

I attribute the occurrence of a seasonal food deficit in Kilimi partly to the difference between the rainforest environment of the Mende and the Guinea savanna environment of the Susu. This difference is illustrated in Figure 3, which compares monthly rainfall in Njala in central Sierra Leone to monthly rainfall in Kilimi in northwest Sierra Leone.[3] The former has a precipitation of 2750 mm per annum and a positive water balance for seven months of the year. Its major ecological hazard is *unexpectedly heavy early rains* that may result in a poor swidden burn in too-wet sites, causing subsequent problems of low site fertility, excessive weed growth, and reduced yields (Richards 1986:101, 1990:268). By contrast, the latter has a more narrow seasonal window for production – an annual rainfall of 1811 mm and a positive water balance for five to seven months – and presents greater attendant difficulties for farmers. Its major hazard is *a period of unreliable precipitation at the end of the rains*, which can result in failure of a crop to mature. Other major hazards include devastating dry season savanna fires, which may enter and prematurely burn new swidden sites, and crop raiding by wildlife, which may necessitate the difficult

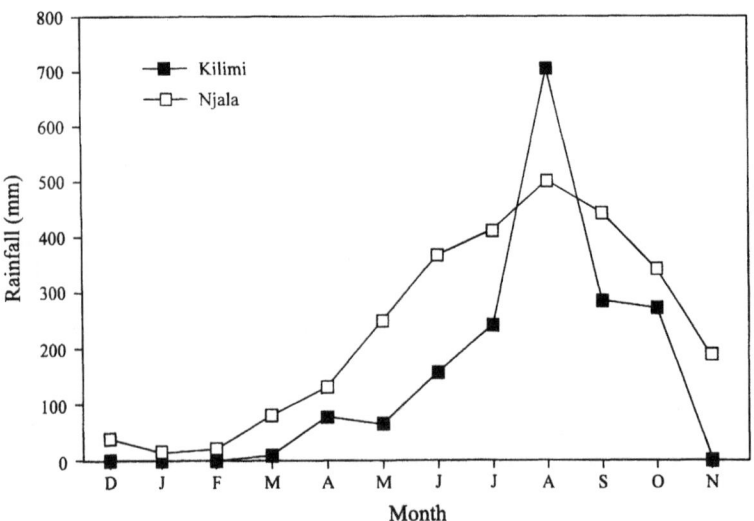

Figure 3. Monthly rainfall in Kilimi and Njala.

hungry-season choice between guarding ripening fields and leaving them unprotected in order to engage in off-farm gathering and fishing activities (see Nyerges 1987, 1988).

The Susu problems of production, however, go beyond an overall food deficit and include a substantial pressure to redistribute both labor and yields. According to some Kilimi farmers, the pressure to redistribute grain inhibits them from planting more of the short-duration rice varieties, such as *nyamaku*, which are agronomically suited to this region. These varieties will ripen during the rainy season and inevitably be consumed directly off the fields by those still awaiting a crop, for example from the common long-duration variety, *samban-konkon*. Most farmers, particularly the elder men with aspirations to political leadership, operate under the expectation that kinsmen and others in need will harvest from their fields, often without asking for permission or stating their intentions. Indeed, so deeply ingrained and pervasive is the customary right to harvest from another's fields that I had great difficulty in returning my harvest samples, once I had threshed and weighed them, to the farmers from whose fields I had taken the grain. I had harvested the rice and therefore, they asserted, "it is yours," and they would not have it back. Finally, the problems of production in this zone also include an extreme variability in yields from farmer to farmer. Examining the source of these yield variations and the general pattern of farmer success and failure from a combined technocentric and sociocentric perspective, that is, from the perspectives of both coping and practice, is the primary goal of the rest of this chapter.

A GERONTOCRATIC AND PATRIARCHAL SOCIETY: THE REDISTRIBUTION OF CROPS AND LABOR

I have argued several times that the key problem in Susu farming relates to the organization and mobilization of labor. In particular, Susu society is characterized by a hierarchical differentiation among farmers in terms of access to and control over sufficient labor to produce and retain adequate yields. In Susu society, inequality in labor organization and control is maintained primarily through poly-gyny in marriage and the system of gerontocratic, patriarchal social control. This system of social and economic control exercised by the elder men over the women and young men results in what amounts to a scramble for available labor. Because of this social environment, some farmers, particularly young men attempting to farm on their own, are unable to mobilize sufficient labor and therefore cannot produce adequate yields. Other farmers, particularly the elder men with greater access to and control over labor, are led to intensify production, grow cash crops, and take greater farming risks in order to maintain their positions. This situation, in turn, has significant consequences for production.

Because of its central importance to the thesis of the ecology of practice, the social system of the Susu bears some further discussion and elaboration. According to my analysis (Nyerges 1992), the key structuring element of Susu society is the competition for social control among elder men, as exemplified in the economy of wealth-in-people and the resulting fusion-fission dynamic of village organization. Thus, as each elder struggles to achieve and maintain political independence from or superiority to his peers, new villages are established and abandoned and village membership changes. In this system, village founders and their descendants are considered to be the landowners or "citizens" of the place, and all others are strangers, newcomers, or slaves. These relationships, however, are fluid and dynamic as landowners use their rights of primacy to recruit outsiders rather than to exclude them, and as villages are regularly established and then abandoned. Above all, as newcomers and slave descendants acquire wealth and position in the community their origins may be renegotiated or conveniently forgotten. Richards describes a comparable, although perhaps exceptionally flexible, Mende system in *Coping with Hunger*,

> descent and history follow the distribution of power and wealth not the other way round. When fortune, misfortune, or clever dealing alter the balance of power and resources within the community, history and descent are adjusted to accommodate the new *realpolitik*. (1986:61)

Richards further provides a metaphor for this flexibility, describing the life of a stranger in Mende society in terms of his conceptualization of a tenure-track assistant professor's position in an American university:

> Strangerhood in Mogbuama is what in American academic parlance would be called a tenure-track appointment. There is a promise of a permanent position, provided the candidate is prepared to work hard and keep to the rules of the system.... (1986:63)

This social fluidity characterizes the Mande-speaking societies of the historically frontier region of Guinea, Sierra Leone, and Liberia. In it, social categories that differentiate between founder and stranger –

between landowner or citizen, on the one hand, and newcomer or slave (or slave descendant), on the other – are permanent features of community life. Individual designation in one or another of these categories, however, is not necessarily permanent. An elder man, whatever his origins, can strive for and perhaps eventually attain suffi-cient wealth-in-people to alter his social standing.

Consequently, it is the careers and lives of individual elder men that form the core of an understanding of Susu ecology. For as much as elder men struggle for political ascendancy over one another, they also try to control other people and their labor, using wealth, ritual links to ancestors, access to Islamic learning, and organizational dominance over land and people to attract and retain adherents, including clients, wives, and junior men. The women and juniors, of course, are not passive actors in this system. Faced with this controlling influence, they inevitably try to escape their obligations and free themselves from the control of the elders "who take too much" (see Bledsoe 1980 for a comparative study in another regional society). In particular, a common strategy for junior men is to alternate between affinal and agnatic residences, or if possible to reside neolocally, thereby anticipa-ting the fissioning process of later life in which elder men aspire to set up villages on their own. An example is the young man called Yana Musa, "Musa from Yana," living away from his natal village near the residence of his widowed sister. A successful farmer in his adopted village, he vigorously defended his move and asserted his intention of returning to Yana once he was better off and more able to cope with the rapacious demands of his elders. He is the source of the junior men's complaint about the elders "who take too much."

Like the junior men, women among the Susu occupy a subordinate social position. Susu gender relations have long been remarked on. Thus, Kenneth Little, writing about research in Sierra Leone conduc-ted in 1945 and 1946, notes that "in the eyes of the Mende . . . Susu are observably more successful in keeping their womenfolk out of the way of other men and under general control" (1951:274). A generalization from more recent research in a neighboring Susu chiefdom, however, that a woman "is expected to be submissive and obedient in the face of all male authority, especially that of her father and, later, of her husband" (Thayer 1981:76), does not adequately characterize the posi-tion of women in Kilimi. Here, women have significant agricultural and domestic responsibilities and in particular may attempt to attain economic independence through groundnut farming on their own. Thus, in Kilimi villages women's economic activity and independent behavior gives them a measure of social significance not fully conveyed by a dominant ideology of male superiority. Their independence is clearly an important element in the structuring of this society, for example in obliging the elder men, despite their advantageous social positions, to adopt social and economic compromises. Nevertheless, the individual freedom of women, as well as of junior males, is constrained by the limited availability of outside social and economic opportunities and by the substantial and strongly determinative control over marriage that is exercised by the elder men.[4] Thus, for a Susu girl, completing her initiation means that she is about to get married, often into the large established household of an elder man. For a Susu boy marriage is long delayed, following initiation in adolescence, as he functions in the work force as a dependent son or prospective son-in-law performing bride-service tasks, i.e., laboring as part of a marriage arrangement in which work instead of wealth is exchanged for a wife.

This unequal distribution of power and social control in Susu society has clear outcomes for household demography. In Kilimi, the polygyny rate ($n = 85$ Kilimi households) is 1.8 wives per husband, a high rate for this trait in the present day and comparable to other Sierra Leonean (i.e., Mende, Temne) examples of the last generation (Little 1948, 1951; Dorjahn 1958). Yet while 50% of household heads have one wife and 26% have two wives, 17% have three, four, or in one case seven wives.[5] The distribution of household sizes is similar. As shown in Figure 4, most households – 81% – fall in the range of one to nine members and account for 58% of the population. The remaining 19% of households fall in the range of 10 to 23 members and account for a disproportionate 42% of the population. These larger households are headed by elders – men whose positions, influence, and power are marked by their polygynous marriages, number of children, and other dependents.

Elder men not only control women's labor through polygyny but also manipulate the members of the young men's cooperative work group, the *simoi*, which constitutes the main source of labor for clearing upland rice fields and planting the rice and intercrops (Nyerges 1988). Thus, Susu elder men attempt to meet the high labor demands of swidden farming through polygyny and by controlling the labor of young men in the cooperative work group. They accomplish this manipulation by having a dependent son, foster son, or client join the group on their behalf; by calling on younger men to repay debts through service; by chiefly rights; or by purchasing the group's labor

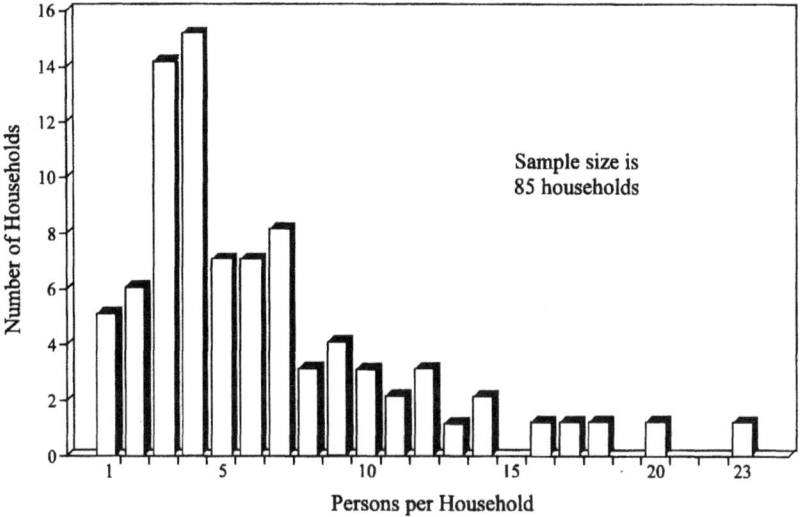

Figure 4. Household size in Kilimi villages.

on a daily basis. For example, of 165 work-group days of the study-village *simoi* recorded in 1982–1983, 55.8% were days employed in working on the farm of a nonmember, that is, a member working for his father, foster father, older brother, or patron (40.6%), sold to an elder by a member or the whole group (7.9%), given by a member to an elder kinsman (4.2%), given by the group to an elder leader or village chief (1.8%), or given to the blacksmith in exchange for smith-ing (1.2%).

The inequality inherent among Susu farming households, reflected in household size and farmer access to and control over farming labor, is further exemplified in the great range of areas cleared for Susu farms. Thus, in a sample of 32 sites measured throughout the Kilimi area in 1983, upland rice farm sizes ranged from 0.15 to 4.09 ha. For the farms in the study village, upland farm size ranged from 0.17 to 3.47 ha and is roughly a linear function of the number of people in the household adult work force. Figure 5, which differentiates farmers between elder and junior men, shows that elders in general have larger households and larger work forces at their disposal, and therefore produce larger farms. In this sample, the smallest farm produced by an elder was that of an occupational specialist – the village hunter who produced a farm of 0.54 ha. When this special case is excluded,

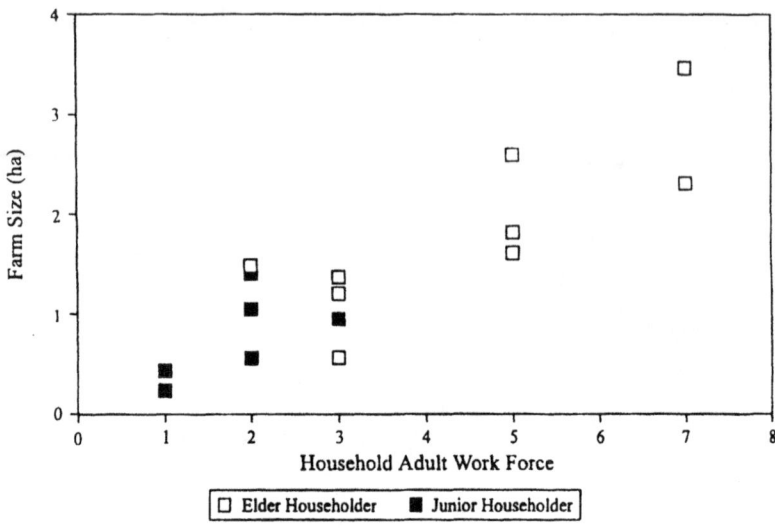

Figure 5. Household work force and farm size.

the remaining elders' farm sites averaged 1.98 ha and ranged from 1.34 to 3.47 ha in size. The juniors, having fewer helpers and incomplete control over their own labor, produced smaller farms that averaged 0.74 ha and ranged from 0.17 to 1.40 ha.

The relationship of yield – measured as weight of rice produced per 10 m^2 sample plot – to farmer status, however, is much more problematic than household and farm size and dependent on other social factors relating to labor, as several examples will suggest. Thus, of 15 upland rice farm sites cleared by men in the study village in 1983, the two smallest (0.17 and 0.40 ha) are farms in which only one person is a worker. These farmers are young men who work part of the time for their elders for bride service and are married only in the sense that they perform bride-service work for a wife. They work on their own farms without any effective aid, either in weeding or in guarding, from their prospective wives. Sample yields from these two farms are 0.25 and 1.45 kg rice/10 m^2 sample plot, reflecting the great difficulty these farmers have in achieving consistent agricultural production. By contrast, a larger farm (1.02 ha) made by a young man permanently married to a wife residing with him, and who in addition possessed an extensive network of non-household helpers, produced a more respectable yield of 2.28 kg/sample plot. Finally, the two largest households (7 adult workers each), which also had two of the largest farms (3.47 and 2.32 ha) and belonged to two of the most prominent men in the village, produced yields ranging from 0 to 2.55 kg/sample plot, for average yields of 1.71 and 0.68 kg/10 m^2 sample for the two farms. The data indicate that in spite of their advantageous social and economic positions, the elder men, much like the junior men with their disadvantageous positions, often experience great difficulty in producing consistently high yields per unit area farmed.

Farmers of the study village also vary substantially in total yields attained from their farms and, equally important, in their individual ability to retain their yields at harvest time. The dimensions of these variations in yields at harvest and in stored yields after harvest can be seen in Figure 2, which differentiates study village farmers on the basis of age (or status) and gender category. This figure shows that, for example, the elders who generally farm larger areas also often produce larger yields than the juniors, although the overlap in yields between the two categories of farmers is substantial. Indeed, this overlap is greater than the overlap in farm size as shown in Figure 5, suggesting that elders are often less successful than might be expected, on the basis of their social standing and access to resources alone, in producing large yields. Figure 2 further shows that the elders on average stored only 10% of their crop, and the portion of the crop stored by these farmers ranged from 1.6 to 25.5% after harvest. The elders can be divided into a group of politically ambitious leaders who, whatever the absolute yields they achieved, redistributed most of their crop to dependents. These elders stored only very small portions – 4.3% in the case of the elder with the largest farm and 1.6% in the case of another ambitious but unsuccessful elder. Another group of politically stable or less ambitious elders were somewhat more successful in holding onto yields, storing between 8.8 and 25.5%.

These variations in production and redistribution can best be understood as the consequences of the effort by elder men to control the labor of women and other (usually junior) men, at the same time as these other members of society are striving to assert their independence. These conflicts for control over persons and their labor result in adopting patterns of labor organization and resource management that may lead to agricultural difficulty. Thus, for all farmers, both household and work-group labor are deployed in the seasonal production of rice and the other crops required for subsistence. In addition, to pay taxes, purchase goods and foodstuffs, and make bridewealth payments for themselves and their dependents, the elder men engage in cash farming, particularly of chili peppers (Nyerges 1988). While both elder and junior men make upland rice subsistence farms, for which they require *simoi* work days, there is intense competition over the control of *simoi* days, and the actual distribution of work-group days represents a net loss of more than half the group's work by the junior farmers to the elders. In this situation, the elder men acquire the labor of the young village men without having to reciprocate directly with expenditure of their own (although they pay for the labor in kind, in cash, or as part of a bride-service arrangement). Yet on days when a work group is not present, most elder men continue to work on their own farms alone. The demands of the economic and agricultural system require that they do this despite the fact that, for an elder man, felling a swidden is personally risky.[6] The elder men's advantageous socioeconomic position does not result in increased leisure for them, but rather, because of the effort to achieve and maintain position as measured through the size and cohesion of the group they control, the elder men are led to work harder and intensify production.

The elder men's adoption of risky rather than cautious management practices, as suggested by their erratic yields, also is the result of the labor shortage in combination with the frontier problem of group cohesion. The risk-taking is a function of the frontier context, because the elder men in this society experience relentless demands for subsistence and cash crop production. In other words, to hold a group together (including being able to pay bridewealth installments to in-laws and thereby retain wives as agricultural laborers), the elder men face the incessant need to redistribute food and money, and must rely on agricultural production to achieve this end (Kopytoff 1987:40–48; Nyerges 1992). More than just the social stratification of the age, gender, and kin group hierarchy, the pattern of manipulating agricultural labor and selectively intensifying production on specific sites reflects the frontier problem of group impermanence. The elder men's advantageous social position, consequently, is not an end in itself, but rather is a tool used to mobilize the labor to produce the crop needed to hold a group together. And far from experiencing increased leisure or greater security through social control, elder men characteristically work harder, store less of their own yield, and take greater personal and economic risks than others.

CROPPING AND FARM SITE CHOICES – THE MENDE AND SUSU COMPARED

A comparison of Susu farming adaptations with the key agro-techni-cal coping strategies of the Sierra Leonean Mende, as analyzed by Richards (1986, 1990), further illustrates the labor and organizational problems inherent in Susu farming and justifies the adoption of a sociocentric perspective in my analysis. According to Richards, the farming adaptations of the Mende include cropping and site choices and the integration of these coping adjustments into a highly flexible social system. Thus, the Mende intercrop rice with short-duration hunger breakers such as maize and long-duration starchy tubers such as cassava. They interplant millet, sorghum, and vegetables with the rice, and plant multiple varieties of rice. They also intercrop cotton, which provides a source of cash when spun and woven into country cloth.

In addition to using the rice farm as a source of multiple food and cash crops, the Mende choose farm sites based on characteristics of site microecology. In his description of the Mende farming system, Richards (1986:28) cites "the crucial importance of topographic plant-ing strategies (following the soil catena [i.e., slope]) in the integration of upland and wetland and the smoothing out of seasonal labour supply bottlenecks." Mende farmers in Richards' Mogbuama study village exploit both moisture-retentive river terrace soils, known as *tumu*, and free-draining upland soils, known as *kotu*. In particular, Mogbuama rice farmers "choose farm sites that follow the soil catena, thus giving them access to some moisture retentive soils but allowing them to plant the greater part of their upland rice on free-draining upland soils" (1990:270). And further, "a catenary farm has consider-able scope for rolling adjustment, allowing for a change of emphasis up or down slope according to conditions as they develop" (Richards 1990:270). This Mende farming strategy allows farmers to take advan-tage of the benefits of moisture retentive *tumu* and flooded sites, where short-duration and floating rices can be planted. Simultaneously, it allows them to plant most of their crop on free-draining *kotu* soils, thereby minimizing the key farming hazard of swiddening in the rain-forest environment, that of early rains leading to poor swidden burns on too-wet sites. Finally, the Mende social system is sufficiently flex-ible that access to and choice between *kotu* and *tumu* soils is equally available to all (Richards 1986:63).

Similarly in the Susu farming system, farmers plant multiple rice varieties, plus maize, sorghum, and millet. Unlike the Mende, however, they cannot intercrop cassava in rice fields distant from a village because they lack the labor to guard year-round against crop raiding by baboons and warthogs. These animals are apparently infrequent and not a farming hazard in central Sierra Leone, but they are numerous in Kilimi where baboons outnumber people by a ratio of 1.3 to 2.5:1. A primatological survey of Kilimi estimated nine troops totalling between 700 and 1350 baboons in the area and reached the conclusion that "baboons are thriving at Kilimi" (Harding 1984:110). This research further shows that

> [t]hese baboons are definitely crop raiders, as is the case elsewhere in Africa, and although the survey team arrived after most of the local crops had been harvested, baboons could still be found around farms in search of gleanings. Concentrations of 50 and more gathered on abandoned upland rice threshing floors in order to forage through the accumulated chaff, and baboons were seen uprooting the few remaining groundnut plants. Local farmers maintain a daily watch against baboons as long as there are crops left to harvest. (Harding 1984:109)

Cassava remains in the ground year-round, exposed to constant pressures of crop raiding, and Susu farmers grow only limited amounts of this crop in garden sites near the houses and close to the village; consequently, it is not a major component of diet. The farmers of Kilimi also do not plant cotton, but instead farm groundnuts and chili peppers as cash crops.

Site choice is also a somewhat different matter for the Susu than for the Mende. In the Kilimi pattern of land use, farmers clear fallows of varying ages and locations and exploit them for varying rotations.[7] The key features of sites recognized as important by Susu farmers are the age of the fallow regrowth and the presence or absence of groundwater in the site. Thus, a key term in Susu agriculture is *ye yire*, literally meaning "wet place," which designates a slope where groundwater is present and that therefore remains moist during rainless periods when other sites turn dry. For example, during the rainy season of 1983 I collected many soil samples from Kilimi farm sites for laboratory analysis. I air dried these samples and discovered large differences between wet and dry weights for samples taken from locations identified as *ye yire*. Some samples from these locations were as much as 30% water, and this water content declined only very slowly as days after rain increased. By contrast, soils from sites not designated as *ye yire* had little or no water content, including one sample collected the day after a 25 mm nighttime rainfall that had only 3% water by weight. Clearly, the moisture retentive *ye yire* soils can significantly buffer the effects of the week-long end-of-the-rains droughts which may occur in this region (see Nyerges 1988) and which spell disaster for farmers who have chosen strictly rainfed sites for planting.

These moisture-retentive, or "phreatic," sites may be ecologically equivalent to the Mende *tumu*. In combination with adjacent uplands where free-draining rainfed, or pluvial, sites are available, these sites provide Susu farmers with site choices comparable to those available to the Mende. But the integration of these sites into the Kilimi farming system is very different. Although recognized as highly productive for growing rice, the *ye yire* sites are also perceived as minimizing the major climatic hazard of Guinea savanna farming, that is, the risk of a crop being lost through failure to mature as a result of soils drying out during end-of-the-rains drought periods. Overall, a *ye yire* with 15 to 30 years of forest regrowth is the optimal new rice farm site in Kilimi. What is surprising is that this is not the dominant farm site type in the region.

Actual swidden site choices made by Kilimi farmers reflect not only the kinds and agronomic quality of sites that are available to them, but also the differential availability and allocation of labor to farmers under the prevailing social system. In this situation, the phreatic, old-fallow sites considered to be the best for upland rice and intercrops of sorghum and bulrush millet are widely scattered and laborious to clear. Farming a large, old fallow requires substantial work, especially by the *simoi*. Although the *simoi* members are junior men, they expend much of their labor in making the large, old-fallow farms belonging to the elder men. The young men also hope to farm in old fallows, where "the bush is strong."[8] However, because labor is scarce they sometimes choose young sites that require less labor to clear,[9] even though the process of clearing these sites may be difficult or unpleasant because of the thorny underbrush. Moreover, the young fallow sites selected by young men tend to be near the village where their mothers or mothers-in-law are planting groundnuts. The junior men have fewer wives than the elders (indeed, they rarely have more than one), and a wife for whom bridewealth or service is incompletely paid may continue to work more for her parent's household than her husband's. Planting the rice swidden in young fallow "farms of convenience" near the women's groundnut farms may be the only way for a young man to avail himself of even minimal help in farming tasks. In these site choices made by junior men, therefore, considerations of agronomic suitability in terms of fallow age and soil moisture may be secondary to considerations of time, labor, and available personnel.

Women control their own purses and, provided they can afford seed, farm groundnuts in their own right. They gain access to groundnut land through their fathers or husbands and may, for example, plant groundnuts in rainfed savanna sites. Alternatively, they may choose to reduce travel time by planting near villages in degraded forest fallow sites. They might also plant in the last year's rice farms. These sites are easy to clear, and the men may want them weeded for the chili pepper cash crop still growing in the field. The women will cooperate in this exploitation of their labor, however, only if the farmer has chosen a relatively dry site, as groundnuts will rot if there is groundwater in the field. That means that in order to intensify production by getting women's cooperation in producing a second-year chili crop, a man must choose pluvial as opposed to phreatic or *ye yire* sites for rice and the other intercrops in the first-year farm. Only in this way will he attract the cooperation of women groundnut farmers in using his site for second-year farms.

The consequence of these labor constraints is that for the elder men the choice between farm sites of different moisture qualities is based as much on social and organizational factors as on cropping and agronomic factors. In the study village in 1983, for example, two-thirds of upland rice swidden land was pluvial, and only one-third was phreatic. This tendency to farm rice in the flat, pluvial sites for the sake of the second-year chili pepper and groundnut rotation puts the better-off elder men at risk if the rains are poor (as happened in the study area in 1983 – see Nyerges 1988:92-93). Examination of data on rice yields by site ecology reveals the lower mean and maximal yields from pluvial as opposed to phreatic sites. Thus, data from the study village in 1983 show that yields of threshed (paddy) rice from pluvial upland sites averaged 1.02 kg per 10 m^2 sample plot and ranged from 0 to 2.55 kg ($n = 21$ samples), whereas yields from phreatic upland sites averaged a higher 1.65 kg per 10 m^2 sample plot and ranged from 0 to 3.10 kg ($n = 10$ samples). Furthermore, the two lowest yields on phreatic sites are from a farm that was neglected during a time of heavy bird depredation. The yields are low, then, due to avian crop raiders (primarily *Quelea erethrops*, the red-headed dioch) and not to problems of site ecology or soil moisture. Calculating rice yield statistics for phreatic sites without these two samples gives a revised mean of 2.05 kg (i.e., double the mean for pluvial sites) and a range of 0.75 to 3.10 kg/10 m^2 sample plot. By contrast, the lowest yields from pluvial sites are clearly due, not to birds, but rather to end-of-the-rains drought and the consequent failure of the rice grains to form in the panicles.

Although the pluvial sites are large and flat and usually have greater total yields than farms on the phreatic slopes, the data emphasize the comparative riskiness of the choice to farm in the rainfed sites. The risk is avoided by men with fewer wives, who lack sufficient labor and therefore, of necessity, forgo the chili cash crop and farm phreatic hillsides, and whose wives plant small groundnut farms in adjacent savanna sites or in the drier hilltop sections. Instead, it is the elder men who select the pluvial over the phreatic sites, despite the risk to subsistence production, because of the weeding requirements of the chili pepper cash intercrop and the need to attract women groundnut farmers to do the labor of weeding in the second-year rotation.

SUBORDINATE MEMBERS OF SOCIETY

The same social factors that result in problematic site choices – the need to secure dependents and to operate in a circumstance of labor constraints – also result in problems and individual variations in yields and in portions of harvests stored for different farmers. Thus, for a patriarch of a family or a chief or aspiring chief of a town or section, the members of the entire social unit are his dependents. In striving to achieve or maintain the goal of acquiring status through controlling wealth-in-people, in the face of overall food shortfalls and severe seasonal periods of hunger, there is no slacking in the elders' effort to produce a crop and redistribute it to dependents in order to maintain the dependent status. Overall, the Susu pattern of environmental exploitation is based on the redistribution of labor and yield at all levels – from the individual to the household to the community at large. This is the social context in which subordinate, as well as dominant, members of society must operate. In the case of the junior males, the effect is primarily their lack of control over their own labor and yields and consequent great difficulty in advancing by accumulating wealth and dependents. Women are affected by their near exclusion from subsistence rice farming except as laborers or in specific "out of system" ways, that is, in which they do virtually all the labor of farming themselves.

Junior Men

Like the elders, the junior men in the study village had great difficulty in controlling yields, averaging retention of less than one-tenth of harvested rice (see Figure 2). Juniors also produced low total yields, often less than 1000 kg per farm. Not only do these junior men operate with virtually no assistance at all, but they are also subject to manipulation by elder men. In order to avoid this manipulation, they may plant in degraded, near-village areas where elder men are unlikely to plant, because the yield will be small, but where the juniors can escape labor-time demands by distancing themselves spatially from the main farming areas occupied by the elder men. There they may also try, usually with poor success, to gain some labor cooperation from women tending nearby groundnut farms. These young men are unable to spend sufficient time on their farms to produce good yields and are mostly valued as laborers, rather than as farmers, in this system.

To illustrate these problems of young men's lives, I need only cite the example of Foday Lai, whose 1983 farm site in the study village in Kilimi resulted in no post-harvest storable yield, a degraded site that had been cut from a too-young fallow, and a below average return of yield to seed planted. Other farmers in the study area in fact did worse, but I focus on Foday Lai because as we stood in the middle of his farm one day he explained to me his farming failure. It was, as he said, the fault of "those elders who are witches here."

Foday, as he knew well himself, was a victim of the social system. His farming activity could be understood only from the perspective of that system. The failure of his farm was not merely the result of poverty, errors of judgment, or a difficult environment, but was social in origin as indicated in the terms of his own explanation. Thus, his farming failure was a result of competition at the level of the whole community, and his errors of judgment were decisions he made as part of an effort to maintain some independence, as a junior male, in a system of gerontocratic and patriarchal social control. Indeed, he had ignored a suggestion by an elder man to farm next to him in an old fallow area distant from the village, deliberately in order to avoid coming under the elder's control.

In addition to direct control by elder's authority, a specific problem faced by juniors is that while they may be "married" they will not have satisfied any of the demands of their parents-in-law and will not have control of their wives as domestic (that is, farm laborer) partners.

The problem of marriage payments and "satisfying" parents-in-law is a pervasive Susu fact of life given their inability to make effective bridewealth payments. Thus, one young man farming in the village is not represented in Figure 2 because although he was an excellent farmer and his yields high, his wife was withdrawn from him at harvest time pending payment. Without his wife he was unable to thresh enough rice to make a payment, and months after harvest (when I left the research area) he was still unable to thresh his rice, had not made a rice store, and had not yet been able to reclaim his wife. In light of the fact that, one after another, every young man of the village "lost" his wife at harvest time, I was able to comprehend the general admiration shown for another young man's skill in negotiation. In marrying a girl of the village, this young man arranged to reside patrilocally, removing her to his own village, rather than remaining to do bride-service farming in the study village near his parents-in-law.

In contrast to juniors like Foday, another young man was an exception, producing a good yield and storing nearly 20% of it. His success was due to a recent change in status, which he took advantage of by behaving in the capacity of an elder. His parents had died, and as the only son he declared himself to be the head of his household of four married sisters. As he informed his sisters' husbands, "you may have satisfied my parents, but now you have to satisfy me." He received much labor cooperation on his farm, as a result, although the cooperation, in turn, cost him a great deal in harvest-time repayment to those who had assisted him. Overall, balancing the social gain made by harvest-time redistribution with the amount stored past harvest, he had done well.

Women

Women in Susu society function as agricultural laborers on their husband's or father's farms, having responsibility, for example, for feeding men's work groups during the dry season and, later, for weeding and guarding fields and harvesting and threshing crops. They are also groundnut farmers, and their impact as social agents stems primarily from the conflict between their value as rainy season workers, in which they perform weeding and guard duties on men's rice farms, and their independent role in groundnut production. Other than these activities, few women engage in rice farming on their own. The one study village woman who planted a small upland rice farm had no labor cooperation and was entirely unsuccessful in this endeavor, achieving no harvest.[10] However, three study village women were able to farm small mono-cropped plots of swamp rice and did extremely well in terms of yields per ha and ratios of yield to seed. These women were also the most successful of all farmers in controlling yield, averaging retention of nearly one-third of estimated yield as stored harvest (Figure 2). They did all the farm work themselves, purchasing the small amounts of seed rice on their own, and, relative to other farmers, accomplished their farming without entailing expensive obligations. Nonetheless, the plots were small and the total amount of rice produced was low. The normal subsistence and cash intercrops cannot be planted in these sites, and they remain a specialized variant in the system, albeit an important one for older women seeking to achieve some measure of economic independence in a system of coercive social control.

TOWARD AN ECOLOGY OF PRACTICE

In the preceeding sections, I have attempted to convey some of the essential conditions of Susu chiefdom ecology. I have argued that a theory of resource management sufficient to explain this system must combine concepts of both adaptation or coping on the one hand, and what might be described as practice, or the ecology of practice, on the other. The ecology of practice, as I have defined it, recognizes that problems of resource insufficiency and environmental decline are produced through individual human activity operating in particular environmental contexts. As farmers, individuals act in systems of "bounded rationality," in which their views and uses of the environment are based on local knowledge and familiarity and a great deal of common sense. But local knowledge and experience represent only one or perhaps limited sets, and not necessarily the best, of numerous possibilities for exploiting the environment. Indeed, the farmers' understanding of the environment is essentially a product of their social organization, as shaped by environmental constraints, in which different individuals have different statuses in society and express these statuses through varying ways of competing for, using, and perceiving the crucial resources available to them. In other words, farmers in chiefdom environments may be variable resource users if for no other reason than because of who they are in age, gender, lineage, and clan hierarchies, as generated out of their interactions with the environment and in competition with one another.

In particular, the subordinate positions of women and junior men in Susu society, taken together, illustrate the basic problem of Susu farming: that inequalities in access to and control over labor, including one's own, are pervasive features of this society that distinguish it from other, more flexible systems. Thus, in Susu agriculture it is undoubtedly true that individuals try to cope with environmental uncertainty through technological and organizational means. At the same time, it is also true that individuals try to achieve socially defined management goals in the context of established and strongly constraining systems of social hierarchy. Not all such systems are equally constraining, however, and what remains to be examined is the extent to which, and why, Susu ecology varies from that of the Mende in terms of the relationship of power to land use.

As compared to the Mogbuama Mende, as described by Richards (1986, 1990), Susu society is not equally flexible. For example, the same flexibility as apparently found in Mogbuama is not as available to Kilimi farmers in the choice among sites of different ecological characteristics. These choices, in fact, are strongly constrained by farmers' social positions, although the issues of founder versus newcomer clan, or even of "citizen" versus slave or slave descendant, are less immediately important than those of age status and gender. Thus, the Susu elder men, in their competition for social place, face severe demands on production that prevent them from making risk-minimizing site choices. They are further led to constrain junior men and women from operating freely as rational decision-makers, obliging them instead to contest for control over their own labor, yields, and site locations.

The juniors in Susu society are caught in a system in which they are valued more for their labor than their productivity, and in the end they are sometimes unable to advance because of the multiple and conflicting demands placed on their time and effort. For them, the overall social context can be characterized as one of competitive poverty and coercive social control. Even the implication of cannibalism, as suggested by the juniors blaming their problems on "those elders who are witches here," is carried through in the characteristic ways in which people talk about one another. For the Susu juniors, age status is social rather than chronological, and for them, there is a real fear of failure to progress, leaving them permanently consigned to the conditions of marginality, powerlessness, and alienation.

This view of juniors in Susu society begs the following question: How is it that seemingly the same social system, as found among both the Susu and the Mende, can under particular conditions lead to vastly different outcomes for individual life courses? In previous work (e.g., Nyerges 1992), I have suggested that Susu society is structured by conditions of a "perpetual frontier," in which social hierarchy, individual mobility, village impermanence, and low population density caused by malnutrition and disease contribute to a heightened social competition for wealth-in-people, with numerous consequences for agriculture and environment. However, it is ultimately perhaps in the somewhat harsher conditions of Kilimi and the Guinea savanna environment of the Susu, as compared to Mogbuama and the coastal rain forest environment of the Mende, that the explanation lies. Conditions of the Guinea savanna that do this are seasonality and low population density, which together produce problems of agricultural labor bottlenecks. Thus, my conclusion is that as environmental conditions worsen, *dominant individuals take advantage of the structuring potentials of their society to solidify their social positions, regardless of the consequences for production*, by limiting the rights of those less powerful and therefore less able to control the resources around them.

Implications

Particularly under harsh environmental conditions of seasonality and low density of population, as are characteristic of the Guinea savanna, deeply sedimented and relatively inflexible social hierarchies may form. Under these conditions, technological changes in development programs need to be carefully introduced, not only to ensure successful integration of the new technologies into existing production systems (see Fyle n.d.), but also to avoid worsening social inequalities by supplying resources that merely enhance established differentiations rather than solve basic problems. Implications for development planning of this essentially sociocentric approach are that, at the very least, goals and expectations for various projects must be realistically predicated on the basis of an adequate understanding of the society that is being subjected to change, including a full and valid assessment of the degree of flexibility that can be expected of its social arrangements. This is not to say that development should proceed only under quasi-egalitarian conditions, which may in fact be rare, but rather that development planning may have to include calculating the costs versus benefits of making changes that, however necessary, risk increased hierarchization, intensified social competition, and the attendant loss or even wastage of crucial local resources.

ACKNOWLEDGMENTS

Field research for this paper was supported in part by funds from the Charles Lindbergh Foundation. Writing was supported by a Summer Faculty Fellowship and two research assistantships provided by the College of Arts and Sciences, University of Kentucky. I thank Caroline Dunn, Barbara Cellarius, and Nina James-Fowler for their substantial editorial assistance.

END NOTES

1. For a specific contrast, Leach (this volume) describes the population density of the Gola Forest Mende she studies as "about 25% lower than the regional average: about 32 persons/km², compared with, for example, 44 persons/km² in Kenema District."

2. Rice yield data were collected in 1983 from the upland (hillside or catenary) and swamp (flooded valley bottom) farms planted by the Kilimi study village farmers. For each of the study village farms, I recorded information on the following: rice variety by local name, site ecology (sites may be pluvial, i.e., strictly rainfed; phreatic, i.e., both rainfed and moist from groundwater; or fluxial, i.e., covered in standing water), patch size (area planted with each variety), weight of rice yield per 10 m² sample plot, patch yield based on extrapolation from sample weight, and weight of seed planted in each patch according to farmers' reports. I also recorded farmers' statements about how much rice they had harvested and bagged (stored) from their farms for post-harvest consumption and the next year's seed. These data, then, indicate both actual farm site production and stored yield following harvest-time consumption and distribution.

3. The climate record from Njala, near Richards' Mogbuama study area, is a 36-year average from 1926–1962, while the Kilimi rainfall data are for one year – 1983. The Kilimi pattern, however, is confirmed by records through the 1980s from the nearby site of Kotor in the Outamba area of Tambaxa Chiefdom.

4. For example, of the 85 censused Kilimi households only one is headed by a woman – a widow nominally remarried to a section chief from across the Kolenten River in Guinea. Contrast this ratio (1:85 or 1.2%) to the four female-headed units in the 98 households (4.1%) censused by Richards among the Mogbuama Mende (1986:51).

5. The remaining 7% of household heads are widowed or unmarried.

6. For example, the elder men typically suffer from orchitis, or elephantiasis of the testicles, due to their lifelong exposure to the filaroidial parasites *Wucheria bancrofti* and *Onchocerca volvulus*, which are prevalent in this

area and transmitted by the bite of mosquitoes and the black fly. In one case, the study village chief was seriously injured and had to be transported to a neighboring chiefdom, hospitialized, and operated on, as a result of genital hemorrhaging brought about by a swiddening accident that occurred while working alone.

7. The main variations in Susu agriculture are as follows: First, the elder men plant first-year rice farms, usually with a large chili pepper intercrop, in rainfed sites in old fallows, often 30 years old or older. Second, the younger men plant one-year rice farms in groundwater-fed sites in old fallows, if possible, or in young fallows if labor is in very short supply. Third, women plant groundnuts in portions of the previous year's rice farms, or occasionally in one-year savanna sites or degraded short-fallow forests near villages. Finally, a very small portion of the forest fallow land may be cultivated in a third year for millet, following first-year rice and second-year groundnuts and chili peppers.

8. The young men generally profess ignorance about the locations of *ye yire*, relying almost exclusively on elder men's knowledge and their own observations of forest regrowth as guides to site choice.

9. Old-fallow farms are felled in two stages of, first, cutting down the smaller-boled trees (4 cm diameter or less) and, second, felling the larger trees. Then, after the initial burn of these swiddens, heavy labor is invested in chopping up debris and reburning it in bonfires. Trees on younger-fallow farms can be cleared in one step, on a single work-group day, and reburning operations are much less arduous.

10. One Kilimi woman who requested rice farmland from her village chief was the only individual I encountered in my entire fieldwork who was flatly turned down for such a request.

REFERENCES

Appadurai, Arjun
    1986 Introduction: Commodities and the Politics of Value. *In* The Social Life of Things: Commodities in Cultural Perspective. Arjun Appadurai, ed. Pp. 3–63. New York: Cambridge University.

Bledsoe, Caroline
    1980 Women and Marriage in Kpelle Society. Stanford CA: Stanford University.

Conklin, Harold C.
    1957 Hanunóo Agriculture: A Report on an Integral System of Shifting Cultivation in the Philippines. FAO Forestry Development Paper No. 12. Rome: Food and Agriculture Organization.

Dorjahn, Vernon R.
 1958 Fertility, Polygyny, and Their Interrelations in Temne Society. American Anthropologist 60:838–860.

Fyle, C. Magbaily
 n.d. Traditional Technology, Rural Development and the Cultural Matrix: The Case of Sierra Leone. *In* History and Socioeconomic Development in Sierra Leone. A Reader by C. Magbaily Fyle, Ch. 1.

Geertz, Clifford
 1963 Agricultural Involution: The Processes of Ecological Change in Indonesia. Berkeley: University of California.

Harding, Robert S. O.
 1984 Primates of the Kilimi Area, Northwest Sierra Leone. Folia Primatologica 42:96–114.

Huss-Ashmore, Rebecca
 1989 Perspectives on the African Food Crisis. *In* African Food Systems in Crisis, Pt. 1: Microperspectives. Rebecca Huss-Ashmore and Solomon H. Katz, eds. New York: Gordon and Breach.

Kopytoff, Igor
 1986 The Cultural Biography of Things: Commoditization as Process. *In* The Social Life of Things: Commodities in Cultural Perspective. Arjun Appadurai, ed. Pp. 64–91. Cambridge: Cambridge University.

 1987 The Internal African Frontier: The Making of African Political Culture. *In* The African Frontier – The Reproduction of Traditional African Societies. Igor Kopytoff, ed. Pp. 3–84. Bloomington: Indiana University.

Little, Kenneth
 1948 The Mende Farming Household. Sociological Review 40:37–55.

 1951 The Mende of Sierra Leone – A West African People in Transition. London: Routledge.

Nyerges, A. Endre
 1982 Pastoralists, Flocks and Vegetation: Processes of Co-adaptation. *In* Desertification and Development: Dryland Ecology in Social Perspective. Brian Spooner and H. S. Mann, eds. Pp. 217–247. London: Academic.

 1987 The Development Potential of the Guinea Savanna: Social and Ecological Constraints in the West African "Middle Belt." *In* Lands at Risk in the Third World: Local-Level Perspectives. Peter D. Little and Michael M Horowitz, with A. Endre Nyerges, eds. Pp. 316–336. Institute for Development Anthropology, Monographs in Development Anthropology. Boulder CO: Westview.

1988   Seasonal Constraints in the Guinea Savanna: Susu Ecology in Sierra Leone. MASCA Research Papers in Science and Archaeology [Special Issue: "Coping with Seasonal Constraints," Rebecca Huss-Ashmore, with John J. Curry and Robert K. Hitchcock, eds.] 5:86–95.

1989   Coppice Swidden Fallows in Tropical Deciduous Forest: Biological, Technological, and Sociocultural Determinants of Secondary Forest Successions. Human Ecology 17(4):379–400.

1992   The Ecology of Wealth-in-People: Agriculture, Settlement, and Society on the Perpetual Frontier. American Anthropologist 94(4):860–881.

Paine, R.
1972   The Herd Management of Lapp Reindeer Pastoralists. *In* Perspectives on Nomadism. William Irons and Neville Dyson-Hudson, eds. Pp. 76–87. Leiden: E. J. Brill.

Richards, Paul
1983   Ecological Change and the Politics of African Land Use. African Studies Review 26(2):1–72.

1986   Coping with Hunger: Hazard and Experiment in an African Rice-Farming System. London: Allen and Unwin.

1990   Local Strategies for Coping with Hunger: Central Sierra Leone and Northern Nigeria Compared. African Affairs 89:265–275.

Spooner, Brian
1982   Ecology in Perspective: The Human Context of Ecological Research. International Social Science Journal 34(3):395–410.

Thayer, James Steel
1981   Religion and Social Organization among a West African Muslim People: The Susu of Sierra Leone. Ph.D. Dissertation, Department of Anthropology, University of Michigan. Ann Arbor MI: University Microfilms.

Vayda, A. P.
1983   Progressive Contextualization: Methods for Research in Human Ecology. Human Ecology 11:265–281.

CHAPTER 7

Toward an African Green Revolution? An Anthropology of Rice Research in Sierra Leone

Paul Richards
Department of Anthropology
University College London and
Working Group in Technology and
Agrarian Development (TAO)
Wageningen Agricultural University

Scientific plant improvement dates from the beginning of the twentieth century. Application of the principles of selection and breeding to the rice plant yielded modest results in the period up until the 1950s. In the following decade, breeders working in Southeast Asia focused their attention on a new plant ideotype – the semi-dwarf rice type.[1] Earlier improved rices out-yielded farmer's selections by 10–20%, but the productivity gains with semi-dwarf types were much greater. In some cases yields were doubled or trebled. The uptake of semi-dwarf varieties by farmers in better-favored environments of South and Southeast Asia was dramatic during the 1960s and early 1970s. The resulting leap in productivity was dubbed the "Green Revolution." Development agencies then set about the task of replicating the technical success of the Green Revolution in other humid tropical regions. Despite concerted attempts to introduce the new rice technologies to small-scale farmers through integrated agricultural development projects in the 1970s and 1980s, success in Africa's traditional rice regions has proved especially elusive. The present chapter, a case study of rice research and its impact in the nodal country of the West African rice region – Sierra Leone – is an attempt to assess some of the strengths and weaknesses of the Green Revolution approach to plant improvement in Africa, and to consider alternatives.

THE PLAN OF THE PRESENT CHAPTER

The present account of rice research and its impact on agriculture in Sierra Leone draws on a much larger, comparative study (in preparation) of rice research in Asia and Africa by Adam Pain (crop scientist), Michael Lipton (economist), and Paul Richards (anthropologist). In this study, Lipton provided overall direction, Pain was responsible for field studies in Sri Lanka, and Richards for field studies in Sierra Leone. The present chapter, in addition to summarizing some of the main substantive findings of the Sierra Leone field study, attempts also to demonstrate the value of anthropological analysis in understanding the trajectory and impact of tropical agricultural science.

The chapter is divided into three main sections and a postscript. The first section approaches the Green Revolution model for rice plant improvement through a careful reading of one of its defining moments – the conference on rice breeding sponsored in 1971 by the International Rice Research Institute (IRRI – conference proceedings published in 1972). Studies of technology *generation* are sometimes vitiated by a concentration on *impact* (Richards and Ruivenkamp 1996). Unintended consequences of the Green Revolution have clouded the picture of technical options on offer when the debate about crop ideotype was fresh. It must also be admitted that not all social scientists writing about the Green Revolution have demonstrated an adequate grasp of the biological intricacies of this debate. Anthropologists, who set so much store by thinking and speaking "as a native," ought to aspire to higher standards in this regard. Analysis of the IRRI conference debates reveals a number of technical paths not taken. With a glance in the direction of the new cognitive anthropology of religion (Boyer 1992; Lawson and McCauley 1990) it is suggested that failure fully to exploit heterodox options in crop breeding technology might be related to the way states of belief are established and maintained within scientific institutions.

The second section outlines attempts to introduce the Green Revolution to Sierra Leone. In the end, it is argued, the Green Revolution failed because of lack of coordination between agricultural development agencies and research. Vigorous efforts were made to encourage farmers to develop wetland areas for rice double-cropping, but the land development work was often deficient, and too few locally adapted semi-dwarf cultivars were available. Skeptical about the applicability of a semi-dwarf strategy in Sierra Leone, rice researchers adopted a double-handed approach, releasing some semi-dwarfs but at the same time pursuing a farmer-first strategy, i.e., selecting taller hardy types more directly in tune with farmers' existing needs.

The third section traces the impact of this two-handed rice research program on farming in Sierra Leone in the 1970s and 1980s. Adoption trends are· assessed through analysis of the rice variety choices of a nationwide sample of farmers in 1987–88. The picture that emerges confirms that there has been little uptake of semi-dwarf varieties, but that breeders' farmer-first choices have played a useful part in strengthening local capacity to adapt to difficult circumstances. Thus, although there was no Green Revolution per se in Sierra Leone in the 1970s and 1980s, rice research during that period was far from negligible in its impact. However, more could now be done to enhance the synergy between breeders and low-resource farmers under the farmer-first scenario.

A postscript to the chapter then seeks to link the material on rice research impacts to the wider political context in Sierra Leone. Attention is drawn to the "residual" character of the local rice farming sector in a national economy dominated by gemstone mining. During the period under analysis politicians gained more from rice imports than from supporting rural livelihood systems. Insurgency, provoked by the perceived corruption associated with the mineral economy, has spread from the Liberian border region during the early 1990s, offering a violent alternative way of life to rural youth disoriented by diamond mining, and now threatens to bring all regular research and rural development activities in the country to a complete halt. If the Green Revolution was a product of the Cold War era, and perhaps flavored by some of its ideological assumptions, what kind of visionary elements, if any, are projected by post-Green Revolution technology possibilities, and what, if anything, would these visionary elements communicate to the youthful post-Cold War warriors of Sierra Leone?

RICE BREEDING: THE STATE OF THE ART IN 1972

An international conference in 1971 at IRRI headquarters in Los Baños, the Philippines, aimed to forge a state-of-the-art synthesis of rice breeding knowledge as the Green Revolution entered its second decade. Advice on "correct citation" in the conference proceedings, published in the following year, makes the level of generality attempted clear from the outset: "International Rice Research Institute. 1972. Rice breeding. Los Baños, Philippines."

The first five papers in a set of proceedings running to 738 pages are historical reviews grouped under the heading "Advances in Rice Breeding," and they seek to document a number of strategic moves in national and international programs that led to the identification, by breeders, of a new plant ideotype as a central focus for their activities. Particular significance attaches to the story of rice breeding in Taiwan (Huang et al. 1972).

Historically, Taiwan was the meeting place of three streams of rice genetic resources: mountain (dryland) varieties probably originating in the Philippines, wetland tropical varieties (sub-species indica) introduced from the mainland of China, and sub-tropical varieties (sub-species japonica) from Japan.

Taiwan was under Japanese colonial rule from 1894 until 1945. Japanese varieties (to suit colonial tastes) grown in Taiwan during the early part of the twentieth century were known as *ponlai* varieties. A number of *ponlai* varieties were short-statured types that responded well to increased amounts of inorganic fertilizer without lodging. They were also photoperiod insensitive. On a tropical island like Taiwan (and unlike in Japan), this meant it was possible to plant the same japonica variety more than once in a single season. Multiple cropping developed early in the colonial period.

Only after 1945 was any attention paid to the improvement of the native indica rice types. Typically, these native wetland types were tall, leafy, photoperiod sensitive, and prone to lodging under fertilization. But they were often more hardy than japonica types, i.e., being blast tolerant and better competitors with weeds, and for this reason were preferred by many small-scale farmers. Breeders sought indica types approximating more closely to the japonica ideotype that had proven successful in multi-cropping systems, selecting for earliness, semi-dwarf aspect, and fertilizer responsiveness, while seeking to retain the hardiness associated with typical native Taiwanese indica varieties.

Prominent among the improved types released during the 1950s were the semi-dwarf indicas such as Taichung Native 1 and Ai-Chueh-Chen. International trials organized by FAO and the International Rice Commission during 1961–63, in Hong Kong, the Philippines, Surinam, and Sierra Leone, established the wide adaptability and promise of these Taiwanese native semi-dwarf varieties. IRRI made use of the same releases in studies to establish the new international plant ideotype for rice. Taiwan's semi-dwarfs were later used in the development of some of the most successful of IRRI's new releases, notably IR8 and other semi-dwarf types. Finally, the message about the merits of Taiwanese hardy semi-dwarf rice varieties suited to double cropping in the tropics reached Africa directly through the activities of 21 Taiwanese agricultural demonstration teams established from 1960 onwards. One of these demonstration teams came to Sierra Leone and established a site based on an inland valley swamp at Mange Bureh, in western Sierra Leone, not far from the Rice Research Station at Rokupr.

One way to characterize the story of the Taiwanese semi-dwarf indicas in a sentence is to say that these rices were the result of blending ideas and practices from the tropical and non-tropical worlds under the stimulus of decolonization. But Taiwan was also a focus for Cold War ideological rivalry. To what extent was the promulgation of the semi-dwarf message, whether as a direct aspect of Taiwanese aid, or in the more generalized variant developed by IRRI, based on objective evidence of the wide adaptability to tropical conditions of these new plant types? Clearly the technology had to work to have impact, but it may not be unreasonable to suspect that the new semi-dwarfs were also supposed to convey a more general ideological message about the way Western (and more especially American) policy makers envisaged the development of the rapidly decolonizing tropical world under non-Communist conditions. Freed from colonial control, Taiwanese breeders had found ways of replicating productivity gains once associated exclusively with the cultivation of non-tropical rices, thereby opening a window for local cultivators onto potentially lucrative international markets without switching entirely to exotic varieties. Seemingly, the Taiwanese semi-dwarf route offered some of the real gains of commercialization while protecting features of local crop types valued by local consumers.

The 1971 conference offers a number of fascinating clues that issues
of ideology and political economy were already at work to disturb the
smooth objectivity of a debate apparently carried out in exclusively
technological terms. This tension is immediately apparent in Robert
F. Chandler's paper on the impact of the improved tropical plant type
on rice yields in South and Southeast Asia (Chandler 1972). As IRRI's
first director (Chandler 1982), Chandler was particularly associated
with establishing the central proposition of the Green Revolution in
rice, that "no advance in recent years has had as great an impact on
the yield potential of rice as that of plant type" (1972:77). He begins
his paper by asserting general acceptance of the proposition that substantial increase in the yield potential of the tropical rice plant resulted from breeding varieties with drastically changed canopy structure.
He immediately notes, however, the paucity of reliable evidence to
demonstrate how much of this yield potential had been realized in
agricultural practice, thus inviting the conclusion that even at the
moment of its triumph the Green Revolution was seen, even by its
architects, as being as much a matter of faith as of fact.

The applicability of the central dogma was tenaciously defended. In
discussion, B. B. Shahi of the Department of Agricultural Research in
Kathmandu, Nepal, poses the following challenge: "The IRRI plant
type definitely needs high doses of nitrogen for its full expression and
yield. But in developing countries where farmers are poor, they cannot
afford to apply even 20 kg/ha N.... [W]here shortage of nitrogen
fertilizer exists, what would be your alternative for the years to come?
Or do you think high yielding varieties without fertilization can give
as high a yield as a local variety?" (IRRI 1972:84).

Chandler insists, in reply, that because "[n]itrogen application pays
handsomely ... the poor farmer can afford to apply nitrogen." But
there is a resonant caveat: "if he can find the money" (IRRI 1972:84).
Shahi's question about an alternative breeding strategy is ignored, as
is the implication of the point that, by definition, poverty is the condition of not being in a position to "find the money." The answer to this
contradiction, embodied in World Bank rural development practice,
was to stick with the ideotype and try and solve the constraint of
poverty through credit.

In a second exchange S. V. S. Shastry of the All-India Coordinated
Rice Improvement Project noted that "[o]ur ... data indicate tall
varieties grown in [the] wet season have a higher yield potential than
those grown in [the] dry season" (IRRI 1972:85). By implication,
semi-dwarf short-season indica varieties might not be the best

available under all conditions. Chandler, in reply, while prepared to concede more investigation was needed, ducked the issue: "I believe Dr. Shastry's remark is only a comment and needs no response" (IRRI 1972:85).

The organizers of the conference invited Norman Borlaug, a pioneer of the Green Revolution in wheat and maize, to present a paper on breeding wheat for high yield, wide adaptation, and disease resistance (Borlaug 1972). Borlaug's paper is an interesting mixture of state-of-the-art discussion of what was then known about durable disease resistance in wheat and a broad-brush treatment of agricultural policy issues overlapping the territory of social science and development studies. In discussion, Borlaug is asked whether there is any chance of one day running out of resistant genes for a particular disease or insect. His answer is curious: "I think there are more resistant genes around. I *have* to be optimistic" (IRRI 1972:591, my emphasis). "Field resistance" (he explains) may be a matter of bringing into concert genes that confer only a low level of protection individually (IRRI 1972:591). This is a rational hypothesis, but, seemingly, not in itself an adequate explanation of why Borlaug felt *compelled* to be optimistic. Light on this point is to be found in the main body of his paper where Borlaug is disarmingly explicit about the need to mix faith and fact in agricultural development. Farmers need to approach the superiority of the new crop ideotypes through demonstration. "Demonstration must be done in the farmer's field," but "unless these changes are *spectacularly* demonstrated by showing what is possible, one cannot put the change across to the farmers" (Borlaug 1972:582, my emphasis). Demonstrations, in short, must be demonstrations of yield *potential*, not examples of what might be achieved in reality. The rational skepticism of the scientist is out of place; on-farm trials are opportunities for farmers to become *believers*. Indeed (according to Borlaug), radical pessimism of the intellect has no place in applied science. In Borlaug's view "[t]he defeatist spirit is the greatest enemy of progress and it persists and is too widespread among scientists" (Borlaug 1972:582–83).

Borlaug is silent on how much this "defeatism" is a necessary and healthy outcome of the radical skepticism, and pluralism, of science. But the IRRI 1971 conference, in the event, demonstrated a much higher level of skeptical plurality than is evident from the rhetorical moments so far analyzed.

Six papers, and an open discussion, contributed to an assessment of prospects for improving upland rice, in which it was clearly recognized that uplands were more variable than lowland environments and that this called for greater emphasis on dynamic plant characteristics in selection. In discussion following a presentation on "Varietal Responses to Some Factors Affecting Production of Upland Rice," D. J. McDonald commented that "[v]arieties are needed that will not only perform well in harsh conditions but will also respond vigorously to increasingly favorable environments at the same or different location" (IRRI 1972:700).

In a thoughtful overview entitled "Prospects for the Future," Lewis Roberts moved on to raise doubts, stating that "[w]hether the new semidwarf varieties ... represent an optimum plant type for all or almost all ecological conditions, as some claim, is still to be definitively determined" (Roberts 1972:717). His own opinion was that the days of the IRRI semi-dwarf ideotype were numbered: "It seems ... that the cream has been skimmed from this avenue of research and that the future breakthroughs in the improvement of rice yields ... will come from other directions. Breeding for increased physiological efficiency within the plant" (e.g., efficiency in water and nutrient uptake, photosynthesis, and respiration) might be the way forward, he suggested (Roberts 1972:717). This insight was accompanied by a shrewd comment on the issue of belief in science. The semi-dwarf episode had brought *drama* into rice research. A shift toward emphasis on breeding for physiological efficiency, however profitable in the long term, would never offer the kinds of "dramatic breakthroughs ... comparable to that which came from altering the plant type" (Roberts 1972:717). Roberts sensibly calls for a scaling down of expectations, but perhaps without fully appreciating the extent to which these expectations were based on faith not fact.

This returns us to a central issue in the present discussion – how, both analytically and substantively, to handle the belief-like elements in programs for technology development (Mackenzie 1992; cf., Lawson and McCauley 1990).

One of the fruits of the recent revival of interest among anthropologists in cognitive studies is to approach science and religion as cultural attainments with common cognitive moorings (Boyer 1992). From this perspective, it would come as no surprise to find that attempts to stir belief are an important aspect of any scientific enterprise. But as in religion so with science, evidence seems to suggest that induction of the mental states analysts label "belief" is in significant measure a social rather than solo process (Douglas and Wildavsky 1982). To understand the "shape," "flavor," and "texture" of any system of scientific beliefs, at a given moment in time, future work will need more fully to explore socio-cognitive interactions in science, i.e., how certain social processes, such as networking, peer reference, and the reactions of wider society, interact with the cognitive capacities of practitioners of science to structure and sustain beliefs (cf., Latour 1993). Similarly, a sociocentric approach is called for in examining issues of why certain beliefs have become hegemonic at particular moments in the history of science and in developing approaches to ensure a greater, and possibly more healthy, plurality of belief within scientific systems.

The point here is not to suggest that there is some external point of reference from which states of true and false consciousness in science can be assessed, but rather to draw attention to the fact that the world of scientific possibilities is always much greater than the hegemonists of the day suppose. It might be helpful to be in a position more reliably to tap this under-utilized potential through focusing attention on the socio-cognitive processes that sustain plurality, as distinct from those that support hegemony. But widely accepted theories of socio-cognitive interaction in science are as yet lacking (but cf., Fujimura 1992). In the interval, it seems useful to consider science from the ethnographic perspective, since ethnography has often proved quite effective in the informal analysis of the social and perceptual elements that make up belief systems. This is the starting point assumed below, in examining what happened when attempts were made to shift the Green Revolution ideotype, and associated belief systems, from Asia to Africa, triggering one of the world's more radically complex cross-cultural projects in global applied science.

It is no criticism of the Green Revolution to detect within it elements of faith and belief or to diagnose the influence of politics. Where there is a problem is when faith, belief, and politics support hegemonic orthodoxy. The present case study is conceived as a contribution to a larger project just beginning (Richards and Ruivenkamp 1996) concerning both the nature of socio-cognitive processes in applied science and the practical significance of pluralism and greater democratic accountability as factors promoting an agricultural science responsive to the new realities of a post-Cold War world.

THE GREEN REVOLUTION IN SIERRA LEONE

As post-1971 rice research moved down the more complex paths mapped out in the IRRI conference, but without the earlier fanfare, rural development agencies were busy converting the standard Southeast Asian Green Revolution developments of the 1960s into "package" form to be applied in African and other rice growing regions. This process was greatly facilitated by a new enthusiasm for poor-positive rural development programs at the World Bank under Robert S. McNamara (1968 onwards). McNamara, apparently reflecting on emerging lessons of the Vietnam War, focused the Bank's attention on the vicious cycle of poverty–debt–backward technology–poverty. Integrated agricultural (or rural) development programs (IADPs) were designed to help the poor break out of this cycle by offering new seed-and-fertilizer packages, cheap credit, and rural infrastructure improvements (Karimu 1981). This process was to be assisted by a post-Green Revolution reorganization of tropical agricultural research, with international centers (IRRI for rice and IITA and WARDA for rice in tropical Africa) carrying out "upstream" (more generalized) aspects of rice research based on the IRRI ideotype, and national agricultural research systems (NARS) concentrating on "downstream" (or essentially adaptive) research. NARS, in theory at least, were intended mainly to screen and release IRRI and IITA rice types, or incorporate these types in locally produced crosses, and then supply foundation seed to multiplication projects working in conjunction with IADPs.

Sierra Leone was an early target of attempts to transfer the Green Revolution in rice to Africa. During the early 1960s, the Taiwanese introduced semi-dwarf varieties to Mange Bureh and demonstrated the technical feasibility of double cropping in inland valley swamps. The Taiwanese model for intensive wetland development was then taken up by FAO and the United States Peace Corps program. Early IRRI releases were screened at the national rice research station at Rokupr in the 1960s. The first World Bank-funded IADP began operations in eastern Sierra Leone in 1973, offering both land development and input-supply loans to farmers interested in cultivation of high-yield rices in inland valley swamps. The German government provided technical assistance to establish a national seed multiplication project in the mid-1970s. By the early 1980s the country hosted a number of IADPs, funded by a range of donors (Karimu and Richards 1981). Among other more varied elements, each IADP had at its heart a near-standard Green Revolution "package" for wetland rice, i.e., development of inland valley swamps and adoption of fertilizer and high-yield varieties.

Theory and practice were very far apart, however. There was too little basic information on inland valley swamp environments to support the major transformations attempted by the IADPs. Development was often attempted in unsuitable swamps and carried out to inadequate standards. Few farmers ever reached the point where two crops could be planted in one year. Many swamps dried out in the dry season and, when project subsidies were withdrawn, quickly reverted to local use, farmers finding maintenance unduly labor intensive and unprofitable.

Use of wetland environments, to supplement upland cultivation of rice, is recorded in Sierra Leone from the nineteenth century onwards (Glanville 1938). Local usage favors tall flood-tolerant rices in the wet season (flooding levels can be very variable) and dry-season cultivation of crops like sweet potato, cassava, and groundnuts on residual moisture. There is little if any evidence that IADP-type interventions had any sustainable impact on this ongoing transformation (Richards 1985). If anything, the reverse is true, since over-enthusiastic development of unsuitable inland valley swamps sometimes so disturbed water regimes that the land became unproductive, even where farmers sought to revert to local cultivation methods.

A particular deficiency was the lack of coordination between national rice research and the agricultural development agencies. After Independence in 1961, politicians were quick to demonstrate a lack of understanding or sympathy for breeding research. In a famous instance, a minister of agriculture of the SLPP government in the 1960s publicly drew unfavorable comparisons between the Taiwanese project at Mange and the Rokupr Rice Research Station, which he characterized as having stood for 50 years producing nothing. In praising the Taiwanese at the expense of the national research team the minister was, in effect, pinning all his hopes on the possibility that the existing Taiwanese selections were well-adapted to Sierra Leonean environments. This proved not to be the case. Chandler visited Rokupr ca. 1967 but, according to participants in meetings at the time, was unimpressed by arguments from national staff that the semi-dwarf ideotype was unsuited to the variable flood conditions met with in most Sierra Leone wetlands. National researchers screened many IRRI accessions, but few showed promise in Sierra Leonean conditions. Three semi-dwarf crosses (released as **ROK 7**, **ROK 11** and **ROK 14**) incorporated Taiwanese/IRRI parentage, but none was widely adopted since they were vulnerable to iron toxicity and poor water management. The Taiwanese semi-dwarf variety **CCA**, a Peace Corps favorite, proved more robust but is vulnerable to blast in Sierra Leonean conditions.

Rokupr had been established, in 1934, as a research station targeting the special problems of the mangrove rice cultivation zone supplying the Freetown market with most of its locally produced rice (Moore-Sieray 1988). Later, work was extended to inland valley swamp and upland environments. In all three environments the local preference was for medium or tall types that were tolerant of weed competition and diseases. Rather than double-cropping, farmers preferred to manage scarce labor by staggering planting and extending the harvest period through half the year by judicious mixture of short-, medium-, and long-duration types planted in both upland and wetland environments or up and down catenary sequences (Richards 1986). In areas where use of wetland increasingly supplemented production on impoverished uplands, farmers especially welcomed the introduction of a mid-1960s Rokupr cross, **CP4**, a tall long-duration variety suited to cultivation in undeveloped inland valley swamps with variable flooding.

During the 1970s the successes among Rokupr releases included a salt-tolerant cross with some local parentage (**ROK 5**), intended for use in the seaward reaches of the mangrove zone, and a further tall longer-duration wetland type (**ROK 10**), suited to natural-flooding mangrove and inland valley swamp wetlands. The 1970s also saw the release of three successful upland varieties, **ROK 3**, **ROK 16**, and **LAC 23** (now **ROK 17**). All three are medium-duration, somewhat fertilizer-responsive, blast-tolerant types selected from local land-race materials. **ROK 16** is, in addition, an awned variety. Awns are said to offer some protection against bird damage, which is one of farmers' main pest problems on uplands in Sierra Leone. **ROK 3** is a dual-purpose variety that also does well in some wetland environments under local management.

ROK 3 and **ROK 16** are selections made by Rokupr breeder (now retired) Gbey Sama Banya. Banya has an especially interesting background that goes some way to explain his success as a breeder in Sierra Leone. Rokupr breeders were never fully convinced of the suitability of the IRRI semi-dwarf strategy in Sierra Leonean conditions. Political skepticism about the relevance of national rice research may actually have served to heighten the mood of critical awareness among national researchers of weaknesses in the Taiwanese/IRRI model. But Banya had additional personal circumstances conducive to intellectual independence. Trained in practical breeding in the colonial period, he had never acquired a degree. Unconvinced that it was a good use of his talents to go back to the classroom, mid-career, he found himself somewhat "outside the loop" when it came to sabbaticals and offers of international employment (G. S. Banya 1989, personal communication). He chose to spend time that might otherwise have been taken on the international circuit at home and in the villages, observing farmer choices at first hand. Both **ROK 3** and **ROK 16** were identified from among local land-race materials collected from his mother's brother's farm in Kailahun in eastern Sierra Leone. As pure-line selections, **ROK 3** and **ROK 16** out-yield the best local types, under farmer management, by margins of ca. 10–20%. Farmers recognize them as related to local types with which they are familiar. This increases their confidence in adopting them, and both types have spread extensively through farmer informal exchange networks. Survey data understate the true importance of these two releases since both readily assume aliases in local languages and reappear as "farmer selections."

Thus, to an extent, in the absence of a string of successful semi-dwarf releases well adapted to local conditions, rural development agencies seeking to foster the Green Revolution in Sierra Leone were making bricks without straw. The Peace Corps and some of the IADPs promoted **CCA** as the basis for introducing double-cropping in well-managed inland valley swamps, but because of its susceptibility to blast in Sierra Leonean conditions this Taiwanese variety was never approved for release by the National Seed Board. Some projects, apparently oblivious of the rules, imported and screened their own semi-dwarf selections, and some seemingly "escaped" into more general use. One such example is "Yifin," named after a remote village in Koranko country and hailed by the Peace Corps as a "local" semi-dwarf when re-discovered in the mid-1980s; it is more probably an exotic survivor from earlier, undocumented activity by volunteers in the region. Tracing the true parentage of these varieties is a near impossibility, since none has gone through any official screening, registration, and release procedures.

Thus the Green Revolution in Sierra Leone was undermined by poor wetland development and poor coordination of research and development. In retrospect, it seems to have been unwise of the World Bank and other donors to have promoted Green Revolution rice-development packages without solid evidence that the country had access to a sufficient supply of robust, locally adapted semi-dwarf rice types and assurance that others were in a well-advanced pre-release stage. Instead of seeking to strengthen national agricultural research, politicians and rural development agencies at times tended to marginalize the national research system. When the lack of adapted semi-dwarf varieties was pointed out to IADP personnel, interviewed in the late 1970s and early 1980s, the response was either to minimize the issue of variety availability, or to talk in terms of obtaining the "right" varieties straight from an international center such as IITA. The volunteer and NGO sector at times introduced varieties suited to their own needs apparently without being aware of the existence of nationally approved procedures to screen and release exotic materials. Meanwhile, national rice researchers continued to pursue the farmer-first option, building upon their experience, and making selection and breeding choices directly related to local realities as seen by farmers. The merit of keeping alive this farmer-first "string" to the plant improvement "bow" has more recently begun to attract some international recognition (Chambers, Pacey, and Thrupp 1989). Twenty-five years ago, Chandler may have considered Shastry's point in 1971 about the superior wet season potential of tall rices a comment requiring no answer, but IRRI itself has now changed its ideotype for rice in regions with the kinds of difficult and varied conditions encountered in Sierra Leone. Given this retrospective international recognition for the kind of low-level pragmatic strategy pursued by G. S. Banya and others at Rokupr from the 1960s, it is time to pronounce the Green Revolution strategy in Sierra Leone dead and buried.

THE IMPACT OF RICE RESEARCH IN SIERRA LEONE

Research Methods

During the 1987–88 farming season, and assisted by colleagues Serrie Kamara, Osman Bah, and Joseph Amara from Njala University College, I undertook a nationwide survey of the impact of modern varieties (MVs) on rice farming in Sierra Leone. Data were collected from 490 farms in ten localities covering three main rice farming ecologies, i.e., forested and forest/savanna-transition uplands and inland valley swamps, the *boli* grasslands, and the coastal mangrove zone. A second survey was undertaken of a sub-sample of 121 farmers contacted in the previous survey to cross-check and follow up on data concerning variety choice and to compile information on socio-economic characteristics of farm households. Results from the first survey were then carefully cross-referenced with data on rice cultivation and variety choice deriving from a longer-term ethnographic study of a rice-growing area in central Sierra Leone (Richards 1986). Additionally, a reconnaissance study was undertaken of a specialized rice growing environment, the riverine grasslands, not covered in the main survey (Richards 1995). The process of rice research in Sierra Leone was studied through examination of documentary sources, formal interviews with rice scientists and rural development professionals, and via participant observation. Rice researchers were keen to gain feedback on their activities, especially through analysis of the farmer germplasm we intended to collect in the field. This collection activity, it was foreseen, would be useful to both parties. The breeding section at Rokupr planned to use the study as an opportunity to add to its accessions of farmer rice varieties and to assess the extent to which MV releases remained pure under farmer management. We, in turn, would benefit from reliable identification of varieties, and, equally important, gain some first-hand participatory experience of screening and selection work.

As already noted, rice researchers at Rokupr pursued a double strategy during the 1970s, selecting and releasing tall durable varieties suited to farmers' existing conditions, as well as some semi-dwarfs for more intensively managed wetlands. Assessing the impact of the "low input" farmer-first strand, it was recognized at the outset, would impose some special demands on fieldwork. If the semi-dwarfs had a major impact, this would be readily recognizable in the field. The impact of the farmer-first strategy would be less easy to discern, since the aim was to provide farmers with varieties that could be readily assimilated within existing farming systems, rather than to pose a dramatic challenge to established practice. Selections from local land races might not even be seen by farmers as MVs, especially where they had been acquired through informal channels and not from projects.

In addition, semi-subsistence rice farming practices under family tenure in Sierra Leone offer severe challenges to quantification. Swidden farmers vary their farm sites from year to year, and plot sizes are subject to variation within a single season as farmers adjust to changes in the expected availability of seed and labor (the two most scarce inputs). In these circumstances, substitutes for yield per unit area measures of farm plot productivity may have to be found (e.g., seed inputs, as in this chapter, or returns to unit of labor, as in Linares, this volume). Moreover, most households either farm rice for subsistence provision alone or for subsistence and sale. Farmers mix varieties of different durations on different land types, partly to spread labor demands, but also to lengthen the harvest season (up to six months in some cases). Farmers often know accurately the amount of seed planted, but they have no means of recalling the amounts of rice harvested for family consumption purposes during the harvest season. Where obligations to kin are repaid by inviting a person to help with the harvest it is not even polite or politic to inquire how much has been taken. Thus, even if a farmer is willing to reveal figures of rice ultimately stored in the rice barn, for survey purposes there is no means for either farmer or researcher to ascertain the relationship between gross and net harvest return [the only alternative, impractical in a survey, is for researchers to take direct measures of yields during harvest time as a component of ethnographic research, e.g., Nyerges, this volume – *Editor*].

A balance has to be struck between the requirements of econometric analysis and the practical possibility of obtaining, in survey conditions, meaningful quantitative data. Survey interviewing was therefore guided by the adoption of a number of practical ground rules. The first was to try and work out what farmers might reasonably know, and be willing to talk about, and not to stray beyond that point. The second was to try and calibrate responses gathered through the "stand in a field and talk" method with data compiled via long-term ethnographic inquiry. The third was to develop some kind of quality control using test-retest procedures, i.e., re-collecting the same items after a three-month interval in order to measure the consistency with which farmers recalled names, dates, and amounts of different varieties planted. Finally, special emphasis was placed, wherever possible, on obtaining seeds from farmers and planting them out under standard observation conditions at Rokupr, in order to facilitate their accurate identification.

Adoption Processes

Farmers reported the screening of new and unfamiliar types, and adoption of satisfactory varieties and the abandonment of less satisfactory varieties, as a regular feature of their farming practice over time. Three data sets are briefly discussed in this section:

● Reported dates of first adoption of varieties planted in 1987–88.
● Patterns of abandonment of less-satisfactory varieties.
● Sources of seed acquisitions.

Farmers could recall year of adoption for 75% of all rices they planted in 1987. Slightly over one-third of all rices recorded in the survey were, in fact, newly adopted by the farmer concerned in 1987, evidence of a high spontaneous rate of innovation in Sierra Leone rice agriculture. The years 1985–86 account for 28.6% of all innovation and 1983–84 a further 16.4%. These figures are consistent with the notion of a turn-over rate of new rices for old, year-in-year-out, of approximately 35–40%.

Table I breaks the 1987 adoptions down into MV, indigenous African (glaberrima) varieties (IVg), and other indigenous (sativa) varieties (IVs). Of all rices, 21.3% were MV, but nearly two-thirds came from non-project sources (evidence that farmers want them, not just that they are available on subsidized terms). As might be expected, since IADPs offer credit in kind, project sources accounted for a somewhat greater proportion by area – average plot sizes of project MV rices were 1.65 bushels of seed planted as compared to an average of 1.46 bu for all rices. Plot sizes for non-project MV plantings ($\bar{x} = 1.32$ bu planted) are below the overall average, pointing either to constraints in supply (e.g., lack of credit) or caution on the part of adopters. Glaberrima (IVg) varieties accounted for 10.6% of all innovations in 1987, constituting a figure close to that for spontaneous, i.e., non-project-subsidized, MV plantings (13.3%). It is sometimes implied that *Oryza glaberrima* is a "residual" from the pre-colonial period and that varieties belonging to this species are on the verge of extinction in West Africa. But here, by contrast, there is clear evidence that farmers continue to experiment with and adopt glaberrima varieties. Some (e.g., var. **DC Kono**) may in fact be recent farmer selections from spontaneous *O. sativa* x *O. glaberrima* crosses (Jusu, personal communication). Glaberrima varieties are often very hardy and continue to perform on poor soils where some sativa varieties would fail. For this reason, they continue to retain a place in local rice-planting schemes, especially in adverse environments in the forest-savanna transition zone in northwestern Sierra Leone.

Table I. Sierra Leone Rice Varieties First Adopted in Year of Survey (1987)

	1987 Adoptions (1987 Plots)	Total Area Planted (bu)	Average Per Plot (bu)
All Varieties	263 (100.0%)	380.9 (100.0%)	1.46
Project MV (modern varieties)	21 (8.0%)	34.7 (9.1%)	1.65
Non-project MV	35 (13.3%)	46.1 (12.1%)	1.32
IVg (*O. glaberrima*)	28 (10.6%)	37.3 (9.8%)	1.33
IVs (*O. sativa*)	179 (68.1%)	262.8 (69.0%)	1.47

Table II provides data, broken down into the same set of categories, for varieties planted in 1987 but adopted at any time over the ten years before the date of survey (i.e., from 1977 to 1986). Adoptions in this category account for 37% of all varieties where date of adoption is known. The pattern is broadly similar to that in Table I, except that the proportion of glaberrimas rises slightly and the proportion of MVs in the sample reduces (from 21.3% to 12.9%).

On the other hand, the average 1987 MV plot size for rices that had been adopted from project sources during the previous ten years, 1977–86, is noticeably larger (2.76 bu) than either the average for 1987 innovations (1.46 bu) or the averages for older non-MV adoptions (IVg – 1.41 bu, IVs – 1.76 bu). Moreover, it is worth noting that plots planted with earlier innovations tend to be larger ($\bar{x} = 1.76$ bu) than plots planted with the current (1987) innovations (see Table I). The same pattern holds good for both MV from non-project sources ($\bar{x} = 1.78$ bu for older innovations, as compared to 1.32 bu for 1987 innovations) and IVg ($\bar{x} = 1.41$ bu, as compared to 1.33 bu). This seems a common-sense finding, that successful innovations breed success and increased future use, whatever the variety in question.

Comparison of the figure for 1987 adoptions of MV (Table I) and 1987 MV plantings from older sources (Table II) seems to hint at a significant rate of attrition in MV use over time, compared to other (non-MV) innovations (the proportion of indigenous varieties – IVg and IVs – remains more or less the same in the mix of current and older innovations). But when the focus is shifted to the (reduced) number of farmers persisting with MVs, the increment for successful innovation seems to be greater than for other (non-MV) innovations. This might suggest that MV-use is somewhat niche-specific, i.e., that not all farmers readily find a place for MV material in their planting

Table II. Sierra Leone Rice Varieties First Adopted in the Years 1977–1986 and Planted in 1987 Survey Plots

	1977–86 Adoptions (1987 Plots)	Total Area Planted (bu)	Average Per Plot (bu)
All Varieties	271 (100.0 %)	478.2 (100.0 %)	1.76
Project MVs	13 (4.8 %)	35.9 (7.5 %)	2.76
Non-project MVs	22 (8.1 %)	39.1 (8.2 %)	1.78
IVg (O. glaberrima)	33 (12.2 %)	46.5 (9.7 %)	1.41
IVs (O. sativa)	203 (74.9 %)	3356.7 (74.6 %)	1.76

repertoire but that where a suitable "slot" is found the advantages are considerable. This notion is consistent with the more general idea that MV material in Sierra Leone contributes to repertoire enhancement, i.e., giving farmers *more choices*, rather than to the replacement of "inferior" land-race materials.

Farmers were asked to name varieties they had abandoned. Ninety-nine farmers cited 167 rice types and reasons why they abandoned them. There were also 24 cases where date and type of rice abandoned could be recalled but not the reason for so doing. Points to note in Table III are as follows. The single largest category of reasons for abandonment of MV and IV material was *poor or inadequate yield* (MV 40%, IV 30%). *Seed/land mismatch* was the second highest reason for abandonment of IV material (18%), but only sixth (7%) for MV. Perhaps farmers gamble more with "familiar" IV material when changing land type, due to rotation of farm land under shifting cultivation, or because of migration, etc. *Pest, disease, weed, or climatic hazards* were the second most significant set of reasons for abandon-

Table III. Reasons Given by Sample Farmers for Abandoning Modern (MV) and Indigenous (IV) Rice Types

	MV N = 45	Percent[a]	IV N = 122	Percent[a]
Abandoned to Try Something Else	1	2%	12	10%
Seed Not Available	3	7%	15	12%
Personal Circumstances (e.g., misfortune)	3	7%	7	6%
Seed/Land Mismatch	3	7%	22	18%
Wrong Duration	7	16%	20	16%
Poor or Inadequate Yield	18	40%	37	30%
Soil Fertility Problems	2	4%	10	8%
Pest, Disease, Weed, or Climatic Hazards	9	20%	17	14%
Plant or Grain Quality Problems	5	11%	14	11%
Processing Difficulties	5	11%	15	13%
Labor Supply Difficulties	3	7%	8	7%

[a]Some rices were abondoned for multiple reasons, column adds to more than 100%.

ment of MV, but these problems ranked only fourth in the case of the (somewhat more hardy?) IV types. In both cases, *wrong duration* (too short, too long) was the third most important reason for changing variety, a factor in about one in six abandonments (16%). *Processing difficulties* and *plant or grain quality problems* appear to be of moderate importance for both MV and IV, but the relative lack of importance of *labor supply difficulties* and *soil fertility problems* in abandonment is worthy of note given their prominence elsewhere in farmers' accounts of the difficulties they face. It is interesting to note that 10% of all IVs were *abandoned to try something else*, i.e., dropped for no negative reason, thereby confirming that speculative innovation occurs among Sierra Leone small-scale "traditional" rice farmers. Consistent with IADP subsidies for distribution of improved seed in Sierra Leone, the problem of *seed not available* was a less important constraint for MV than for IV. Seed availability and *personal circumstances* together accounted for less than one-fifth of all reasons for abandonment (IV 18%, MV 14%). This supports the conclusion that seed character and agro-economic rationality are the most important factors in determining seed choices under Sierra Leonean conditions (cf., Adesina and Zinnah 1993).[2]

Where retained seed is insufficient, most farmers rely upon the local informal seed system rather than purchase to supply their needs. Most of this local seed is acquired on exchange, gift, or loan terms from friends, kin, or richer farmers, and seed loans are one major way in which farmers are introduced to new seed types (Figure 1). But not all seed acquisition occurs in the context of seed shortage. Farmers also regularly exchange small amounts of interesting new or unfamiliar planting material and multiply the seed on farm sites until they have enough for an entire plot or section of a farm. They also select and multiply rogues and adventitious introductions (seeds scattered by birds or elephants, for example), and social solidarity is reinforced by handing on any such new and interesting material to friends and kin.

In the 1987–88 survey, farmers were asked, for each of the varieties currently planted, how they had *first* acquired the seed type in question (Figures 1, 2 and 3). Local farmer-to-farmer exchange was the single most important source of new planting material (30% overall – Figure 2, but 42% for IV – Figure 3). Farmers looking for a specific seed type will offer to exchange rice earmarked for consumption for the seed they desire. In other cases, a friend or neighbor might be given enough interesting material to organize a little on-farm trial in the expectation, on the part of the giver, of receiving something inter-

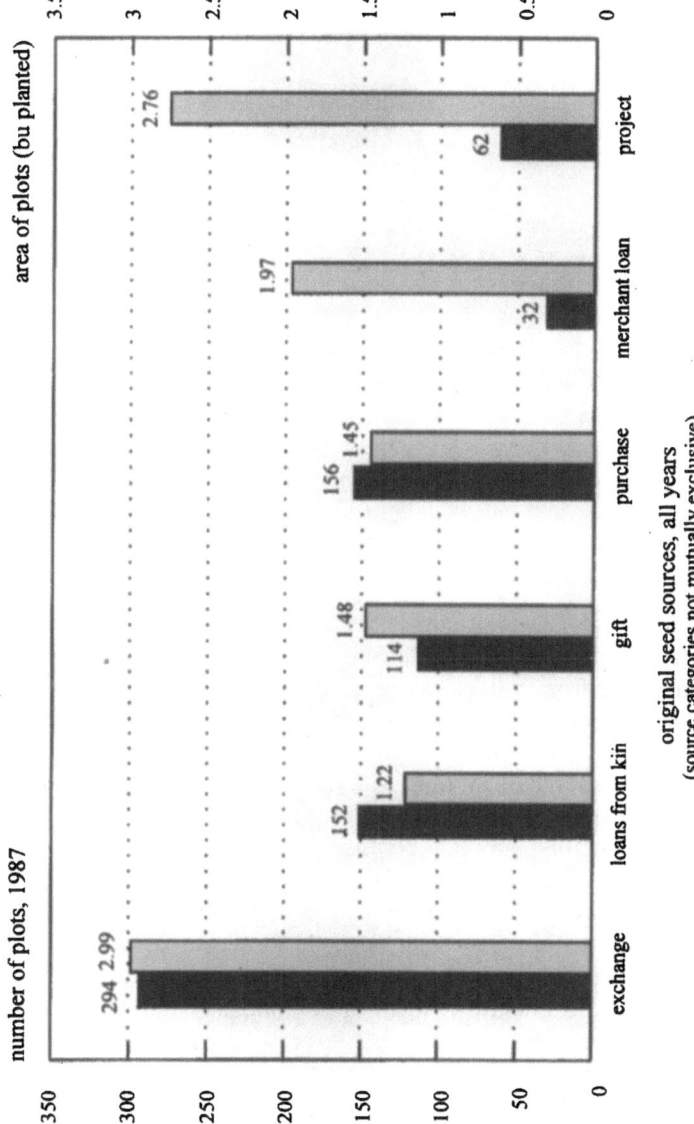

Figure 1. Rice plots and seed sources: number of plots and plot sizes (measured as bushels of rice planted) by original seed source.

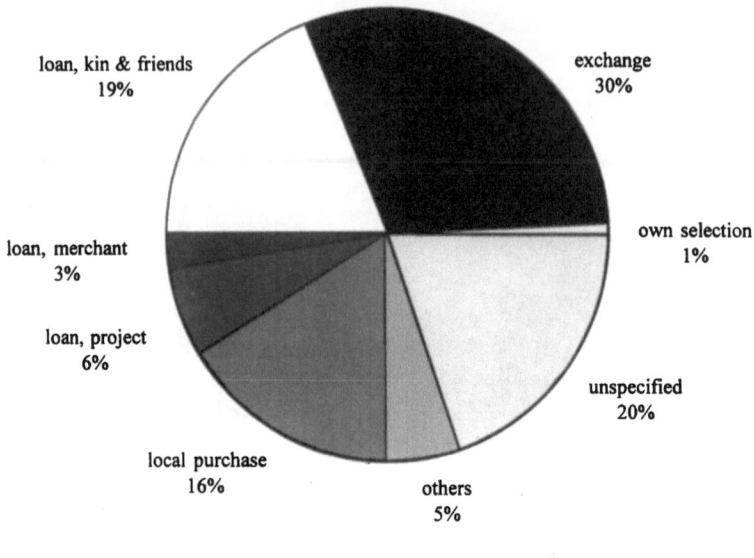

National Sample

Figure 2. Rice seed: all acquisitions by original source for 1987 plantings.

esting in return in the fullness of time. Seed loans from kin and friends
and local purchase from village rice merchants were the second and
third most important sources of new seed types (19% and 16% respectively – Figure 2). It is interesting to note that local purchase was
a more important source of IV than MV (22% of all IV acquisitions,
10% of all MV acquisitions – Figure 3). But if fewer farmers buy MV
than IV materials this may be because MV types are available
through project loans. Altogether project loans accounted for one-
third of MV acquisitions (6% of all seed acquisitions). That two-thirds
of MV material is acquired through informal farmer-to-farmer loan,
exchange, or local purchase (Figure 3) is impressive evidence that MV
choice has become an integral element in the indigenous agricultural
system. It further provides evidence that farmers find breeders' selections useful, independent of IADP subsidy effects. Figures 2 and 3 also
show that farmer selection accounts for about 1% of new seed types
acquired. This appears small, but it may not be insignificant if farmer
selection has cumulative genetic impact over time. The topic is currently under investigation by Rokupr breeders (Jusu 1995).

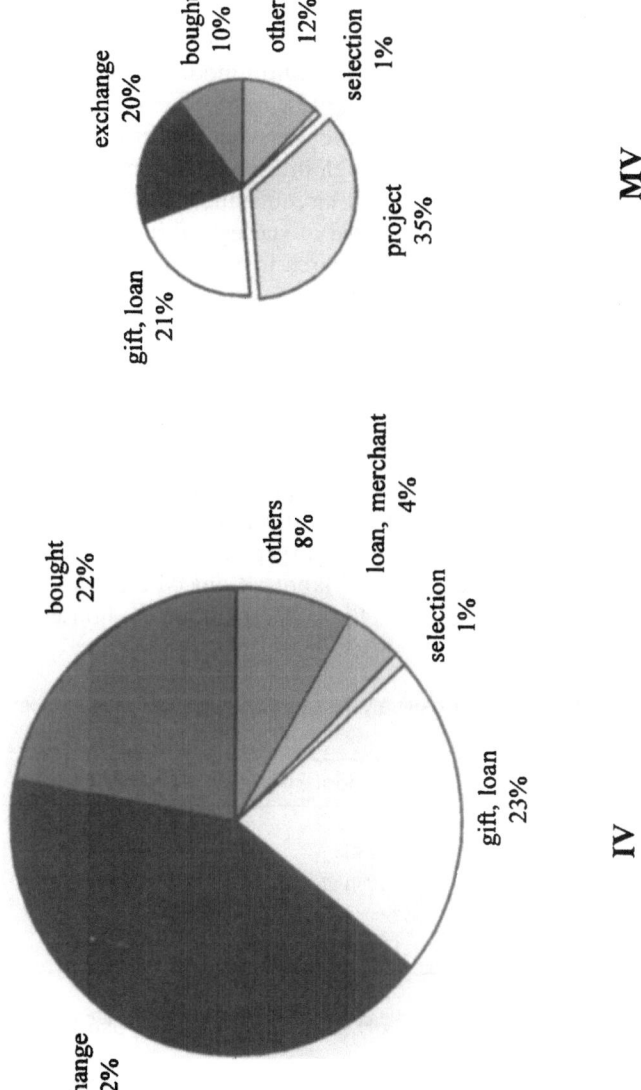

Figure 3. Rice seed: a comparison of IV and MV acquisitions by original source for 1987 plantings.

A final point to be emphasized is that adoption and abandonment of MVs is hardly separable from farmers' more general procedures for assessing, accepting, and rejecting planting material. In Sierra Leone, MVs have had to find their place within an ongoing process of on-farm adaptation. Where the Green Revolution model for agricultural transformation has been abandoned, research systems should be judged according to their ability to operate synergistically in farmer-first (non-semi-dwarf) settings. Although precise measurements of performance are hard to devise in such circumstances, the data below indicate that some considerable degree of success must be attributed to the farmer-first tendency of rice research in Sierra Leone.

Social Correlates of the Adoption of New Rices

There is much evidence from Asia that richer farmers adopt MVs first and benefit more from such adoption. The evidence is less clear-cut for MVs in Sierra Leone. Access to labor, rather than to land, is the main characteristic defining relative wealth and poverty at village level, and Tables IVa to IVc show that MV users did indeed, overall, tend to have more family and hired labor and larger net output of rice. But the bimodal pattern of access to hired labor among MV users should be noted. MV adoption is nearly as high (39%) among the households

Table IVa. Social Indicators Positively Associated with MV Use in Sierra Leone: Hired Labor Supply

HIRED LABOR	MV Adopter	IV User Only
Less than 50 Person – Days/Year	18 (39%)	44 (59%)
50–99 Person–Days/Year	9 (20%)	19 (25%)
More than 100 Person– Days/Year	19 (41%)	12 (16%)

Table IVb. Social Indicators Positively Associated with MV Use in Sierra Leone: Family Labor Supply

FAMILY LABOR	MV Adopter	IV User Only
0–3 Persons	12 (26%)	32 (43%)
4–6 Persons	16 (35%)	21 (28%)
More than 7 Persons	18 (39%)	22 (29%)

Table IVc. Social Indicators Positively Associated with MV Use in Sierra Leone: Disposable Rice Output Per Household

RICE OUTPUT/HOUSEHOLD	MV Adopter	IV User Only
Less Than 21 bu	8 (17%)	35 (47%)
21–39 bu	8 (17%)	18 (24%)
More Than 40 bu	17 (37%)	12 (16%)
No Data	13 (28%)	10 (13%)

with least access to hired labor as among the households with the most (41%). Follow-up inquiry suggested that whereas the larger households, better known to project authorities, may have somewhat privileged access to project-distributed seed, heads of such households, in a political culture dominated by patrimonial values (cf., Richards 1986, 1996), regularly redistribute some portion of MV seed to poorer clients.

According to the survey data, MV users are partly differentiated from non-MV users by age and schooling, the direction of the data in Tables Va and Vb suggesting a slight negative association between youth (and non-Koranic schooling) and MV use. MV users are hardly to be differentiated from non-MV users by any other socio-economic characteristic, such as gender or type of housing construction (Tables Vc–Ve). Generally, there is nothing to prevent most farmers, except

Table Va. Social Characteristics Not Associated, or Negatively Associated, with MV Use in Sierra Leone: Age

FARMER	MV Adopter	IV User Only
Old	10 (50%)	29 (39%)
Middle Age	23 (50%)	29 (39%)
Young	2 (4%)	13 (17%)
No Data	11 (24%)	16 (21%)

Table Vb. Social Characteristics Not Associated, or Negatively Associated, with MV Use in Sierra Leone: Schooling

FARMER	MV Adopter	IV User Only
Schooling	2 (4%)	12 (16%)
Koranic	13 (28%)	18 (24%)
No Formal Education	31 (67%)	45 (60%)

Table Vc. Social Characteristics Not Associated with
MV Use in Sierra Leone: Gender

FARMER	MV Adopter	IV User Only
Male	39 (85%)	65 (87%)
Female	7 (15%)	10 (13%)

Table Vd. Social Indicators Not Associated with MV
Use in Sierra Leone: Type of Housing Construction (I)

FARMER	MV Adopter	IV User Only
Metal Roof	37 (80%)	67 (89%)
Thatch Roof	3 (7%)	4 (5%)
No Data	6 (13%)	4 (5%)

Table Ve. Social Indicators Not Associated with MV
Use in Sierra Leone: Type of Housing Construction (II)

FARMER	MV Adopter	IV User Only
Mud Walls	32 (70%)	53 (71%)
Cement Walls	9 (20%)	19 (25%)
No Data	5 (11%)	2 (3%)

perhaps the young men, from begging a handful of desirable MV from
a patron or friend and multiplying it for themselves. Few of the widely
distributed MVs require any substantial additional inputs. For these
reasons it seems not surprising that the longer-term pattern of MV
usage in Sierra Leone, once a variety is established, is determined
essentially by the agronomic performance of the variety in question.
This is the conclusion of a study by Adesina and Zinnah (1993), and it
is supported by what farmers told us about how they viewed their
seed choices, including MV successes. Evidence for these largely technical rationalizations is presented in Tables VI–VIII and in the notes
of farmers' remarks accompanying these tables.

Farmer-First Breeding: Ethnographic Evidence

The idea that the pattern of MV usage is better explained by seed
characteristics rather than by social characteristics of farmers is supported by detailed work on seed choice processes in central Sierra

Table VI. Mangrove Rices

VARIETY	Plots	Duration (Days to 50% Ripening)	Height (Culm Length, cm)
Kolma	44	153–192 $\bar{x} = 164$ median = 155, $n = 5$	141–150 $\bar{x} = 147$
Soro (various)	35	156–195 $\bar{x} = 166$ median = 162, $n = 9$	101–140 $\bar{x} = 128$
ROK 10	20	161–176 $\bar{x} = 165$ median = 165, $n = 11$	126–145 $\bar{x} = 136$
Fatu yando	20	155–165 $\bar{x} = 158$ median = 157, $n = 7$	141–160 $\bar{x} = 158$
Gbassnin	11	153–178 $\bar{x} = 165$ median = 163, $n = 7$	103–151 $\bar{x} = 129$
Milikwi	11	142–165 $\bar{x} = 153$ median = 154, $n = 8$	97–104 $\bar{x} = 101$

Notes

Six main rices, out of ca. 40–45 distinct types, account for 57% of all plots (141 out of 248) sampled in the mangrove zone. These are among the most successful varieties in the long–season section of the Great Scarcies mangrove environment. The five IVs were generally commended for heavy yield, big grains, and nutritional satisfaction, but their red skin causes women extra difficulty when cleaning them by hand. **ROK 10** was commended for its white skin, but it takes 5–10 days longer in ripening than its main rivals (especially **kolma** and **milikwi**), and this was considered a disadvantage. Some farmers noted that tall varieties (**ROK 10**, 136 cm., **kolma**, 147 cm; and **fatu yando**, 158 cm.) were advantageous in deep flooding areas. Others complained that very tall var-ieties (notably **fatu yando**) were too tall for children and some young women to harvest easily. **Milikwi** is commended for yield, good cooking properties, and as a quick ripen-ing (i.e., "four-month") "hunger-breaker" variety. With some samples ripening in ca. 140–150 days ($\bar{x} = 153$ days; median = 154 days) **milikwi** was the quickest of all cul-tivars sampled in the mangrove zone, and it may be quicker ripening than **ROK 5**. The latter is advertised as a 135–140 day variety, but a sample collected on the Great Scarcies took 170 days. **ROK 10** is valued for its high yield, good taste, and its suitabil-ity for deep-flooded areas, but it is slow-ripening. Shattering of panicles and a tendency for grains to break in milling are deprecated. **Fatu yando** is high yielding, but some farmers find it difficult to thresh. It tends to lodge if not harvested on time but is said not to shatter. **Kolma** is high yielding and is considered to be especially filling – farmers frequently commented that it needed fewer cups of dry rice than average to make a family meal. Earliness, lack of shattering, thin husk, the fact that it is easy to thresh, and that it stores well and is resistant to weevils are appreciated. Some farmers remarked that the panicles tend to lie in the same direction for harvesting. One farmer noted that it is not well-suited to the grassy associated swamps behind the main mangrove areas.

Table VII. Boliland Rices.

VARIETY	Plots (Area in Acres)	Duration (Days to 50% Ripening)	Height (Culm Length, cm)
CP4	15 (24.35)	143–195 $\bar{x} = 180.3$ median = 181, $n = 9$	126.0–145.6 $\bar{x} = 134.0$ median = 134.8, $n = 9$
Mara	10 (9.90)	no data	no data
Thura	6 (9.65)	165–165 $\bar{x}$, median = 165 $n = 2$	142.0–165 $\bar{x}$, median = 148.8 $n = 2$
Gbendembu	5 (7.70)	162–163 $\bar{x} = 162.3$ median = 162 $n = 3$	118.8–121.4 $\bar{x} = 120.4$ median = 121.0 $n = 3$
90–11	5 (5.80)	191–195 $\bar{x} = 193.3$ median = 194, $n = 3$	no data
Rep	4 (7.00)	no data	no data
Indochine Blanc	3 (11.15)	no data	no data
Kobbo	3 (3.65)	no data	no data

Notes
Eight main rices account for 48.4% of boliland plots (45 out of 93 plots) and 52.7% of the sampled area planted to rices (64.4 out of 122.1 acres). Farmers growing **CP4** reported high yields and easy threshing and hulling. Weed tolerance is good when planted on good soil because **CP4** grows high above the weeds, especially if it is planted early. The main disadvantages are that it needs fertilizer, particularly on short-fallow soils, and extra labor is needed to scare birds, especially in the nursery. Milling product is good provided that fertilizer has been applied; 1 threepence pan produces 3 to 4 buttercups of clean rice. White grain makes pounding easy. **CP4** swells up when cooked, but it needs a lot of water. Three cups will feed six people and nutritional satisfaction is high. One farmer claimed that if **CP4** is eaten in the morning, it will satisfy you for the whole day. **Mara** is a quick variety (4–5 months), palatable, and good for feeding large households (it swells when cooked). It produces high yields, although one farmer said that it was not as good tasting or as high yielding as some others. **Mara** is tall in good soil, but careful weeding is needed in poor soil. Pests, particularly birds, are a problem because it ripens early and the grain is white and "sweet." One farmer reported that **mara** matures before cattle arrive in the bolilands, but fencing is required to protect the rice from rodents. This rice needs moisture but does not do well in deep water. One farmer abandoned it after three years because there was too much water in the swamp. A second farmer abandoned **mara** to try a better variety. **Thura** is said to do well in open boli grasslands. It is described as a short duration (5–6 months), high-yielding variety not requiring fertilizer. The milling product is better than average. It is easy to thresh and hull and is bulky and satisfying when eaten. Pests, particularly birds, are a problem on the farm. The rice needs water, especially if the soil is poor. Farmers report contrasting opinions about its weed tolerance. **Gbendembu** is high yielding provided that it is not left late in the nursery. Farmers report heavy tillering and little wastage at harvest, but one farmer abandoned it because of many empty husks and low yields. No fertilizer is needed. Farmers said that cane rats and weaver birds were a big problem.

Table VIII. Upland and Inland Valley Swamp Rices in Northwestern and Northcentral Sierra Leone.

VARIETY	Plots (Area in Acres)	Duration (Days to 50% Ripening)	Height (Culm Length, cm)
DC Kono	29 (68.65)	110–138, $\bar{x}$ = 124.9 median = 126, n = 10	81.4–106.2 $\bar{x}$ = 93.7, n = 9
ROK3	22 (53.0)	126–133, $\bar{x}$ = 130.0 median = 130.5, n = 4	103.8–117.0 $\bar{x}$ = 108.6, n = 3
Gbassnin	20 (24.8)	154–163 $\bar{x}$, median = 158.5, n = 2	117.8, n = 1
Kamara	18 (21.7)	111–140, $\bar{x}$ = 127.5 median = 131, n = 15	73.0–124.4 $\bar{x}$ = 99.2, n = 16
IDA	17 (30.8)	110–144, $\bar{x}$ = 133.4 median = 134, n = 9	83.2–121.0 $\bar{x}$ = 101.5, n = 13
BOROKOLAN	14 (30.4)	110–138, $\bar{x}$ = 124.9 median = 126, n = 10	81.4–106.2 $\bar{x}$ = 93.7, n = 9
Kawulaka	10 (7.81)	130, n = 1	150.6, n = 1
ROK 4	10 (6.9)	123–137, $\bar{x}$ = 127.5 median = 125, n = 4	115.2–125.0 $\bar{x}$ = 120.7, n = 3

Notes
Eight rice types account for 47.8% of plots (140 of 293 plots) and 45.0% of area planted to rices (243.9 out of 541.9 acres). **DC Kono** is a glaberrima type, growing well on poor soils where sativa types fail. This short-statured rice is drought tolerant but is easily overshadowed by taller weeds. It attracts weaver birds. Red grains make it difficult to clean. It is

Table VII. *Notes (Continued).*

Threshing, hulling, and cooking are easy, and the rice tastes good. The white grains swell in cooking and are glutinous and satisfying. Three cups will feed six people. **90–11** grows vigorously and is high yielding. It is a tall variety and "covers" weeds. Farmers describe it as a long-duration rice needing a lot of water. It is very attractive to birds since the grain is white and shiny and does not ripen early. The milling product is poor, though it threshes well and is easily hulled. It is easy to cook except when parboiled and stays firm when stored overnight. It is very "sweet" and feels heavy in the stomach. It is said to be good for large groups. Four farmers in the sample had abandoned **90–11**, two because of its long duration, one because it burnt out in the nursery, and one because it shattered too much when ripe. **Rep** is high yielding if fertilizer is applied, but one farmer complained that fertilizer is hard to acquire. One bushel planted in July will yield 24 bushels in January. Hulling is easy. Birds do not see the grains easily because they are dark. The rice competes well with weeds and produces few empty husks. However, one farmer reported poor yields and complained that seeds were small. Others considered it disadvantageous that **rep** is a long-duration variety and needs much water. One farmer said that **Indochine blanc** must be planted in the rainy season to avoid cattle damage. It is a long-duration, high-yielding varity with good weed competitiveness. It withstands floods by "floating" (i.e., rapid growth of the first internode). One farmer said that shattering can be reduced by spraying water on the crop before the harvest. The grain is white, "sweet," and filling.

Leone, 1982–1992 (Richards 1986, 1995). Ethnographic evidence shows that capacity to assess seed choices from a technical point of view is common knowledge among farming populations in central Sierra Leone.

Mogbuama (population ca. 550) specializes in rice production in the scarp zone of central Sierra Leone. Seed selection studies were undertaken, as part of a long-term ethnographic inquiry, in 1983 and repeated in 1987.

A complete inventory was made of all rices grown on household farms in 1983, and samples of all major types were collected. Seventy

---

Table VIII. *Notes* (*Continued*).

considered satisfying to eat, but some complained that "over-eating" caused constipation. Rated only 2 on a 5-point scale for ease of cooking, the grains tend to swell but remain slightly hard. Cooked rice keeps well. **DC Kono** came from the Kono/Tonkolili region, reaching Masongbala Chiefdom in 1985. **ROK 3** is considered heavy tillering and high yielding. It breaks easily in milling. It is "not very labor intensive," since it is tall and outcompetes weeds. White-grained, it is also easy to clean. It is vulnerable to birds because it is late-maturing and other upland rices have been harvested. It swells in cooking and tastes "sweet." Good for feeding workers and large families, it can be kept for as long as two days after cooking. **Gbassnin** is tasy, heavy in the stomach, aromatic, and a rice that keeps well after cooking. A long-duration high-yielding type (5–7 month), it is easy to thresh. The only two disadvantages mentioned were that its small grains are liable to bird attack, and it requires heavy rainfall. **Kamara** (a glaberrima) is liked for its yields and resistance to bird damage (possessing a sharp apiculus and tough husk). It can be planted anywhere, even in young bush, and competes strongly with weeds, but it wilts easily in drought. Milling product is good, although the grains break if not properly dry. The red bran is difficult to remove. One farmer commented "it is so palatable that is satisfies people readily." It swells in cooking and is heavy in the stomach. The MV "**IDA**" outyields local varieties but needs fertilizer. Yields in young bush are poor. A 5–6 month variety, **IDA** takes long to germinate after sowing, but it tillers abundantly and produces large grains. Bird damage is low on account of grain size, but rodents tend to be a problem. Tall and quick growing, the rice "pushes weeds aside" on fertile soil. Drought tolerance is poor – the rice needs water at planting and flowering. There is little grain breakage at milling and little wastage in threshing. The white grains are easy to clean. (But one farmer reported red grains.) **IDA** swells in cooking and is pleasant to eat. It is filling and useful when a large group has to be fed. **Borokolan** is better to eat than **IDA** because it is "oily." The rice swells in cooking. It is satisfying and provides much energy. Cane rats attack it more than **IDA**. It is high yielding. Early plantings sometimes failed to germinate after drought. **Kawulaka** is a well-liked three-to-four month variety, noted especially for the satisfaction it gives in eating. It cooks quickly and turns out soft (in contrast to **DC Kono**), but it needs a lot of water in the pot. It does not become sticky and keeps well after cooking. Birds also like its sweet taste, but spikes on the grains offer some resistance to bird attack. It needs good soil where it grows to chest height. It is easily threshed and cleaned (white grains). **ROK 4** is high-yielding but needs fertilizer. It is liable to rodent damage unless planted in a water-logged area. It is drought-sensitive but competes well with weeds where soil is well turned. Early maturing, it suffers bird damage. Milling product is high – one threepence pan gives three buttercups of clean rice. The pericarp is white and easy to clean. Grains are small but swell up when cooked, making the rice filling. Three cups of rice will feed seven adults. The fine color is said to resemble "Chinese rice."

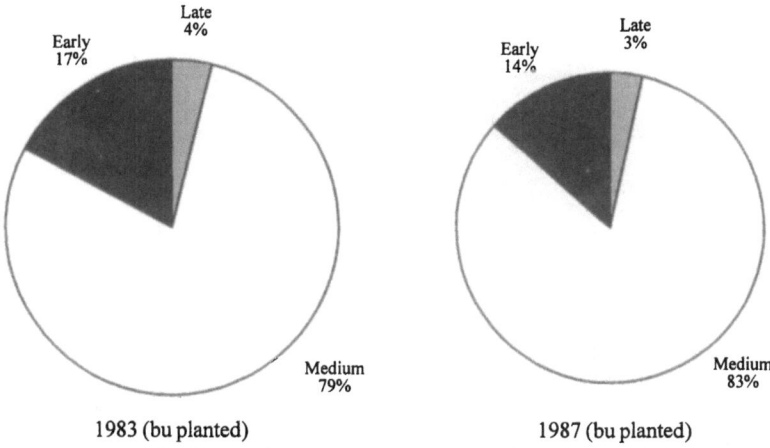

Figure 4. Mogbuama rices in 1983 and 1987: a comparison of early, medium, and late ripening varieties for 52 farmers working in both years.

percent of household farms were inventoried in 1987, and rices were again collected to assess change in rice seed usage over a five-year interval. Samples collected in 1987 were planted in observation plots at Rokupr Rice Research Station in 1988 under the supervision of a breeder, Malcolm Jusu (whose cooperation, and that of the station director, Dr. Sahr Sama Monde, is here gratefully acknowledged).

The general pattern of variety use was unchanged between 1983 and 1987 (Figure 4). About half of all household farms planted medium-ripening (i.e., four-month) upland rices exclusively, although often in catenary sequences with early plantings on lower slopes. The balance of households, with better labor supply, planted, additionally, early ripening (i.e., three-month) varieties on lower slopes and late-ripening (i.e., five-month) varieties in swamps. In both years, late-ripening (five-month, swamp) varieties accounted for between 3 and 5%, and early ripening (three-month, lower slope) varieties for between 12 and 15%, of all rices planted as measured in terms of planted amount of seed (Figure 4). About 40% of rices were planted on water-retentive river terrace soils and 60% in the zone of free-draining gravelly soils.

Within this stable general framework, however, considerable change in variety usage was recorded (Figures 5–7). Some of this change was accounted for by the aging and migration of household heads. Roughly one-third of all heads of farming households had migrated, died, or

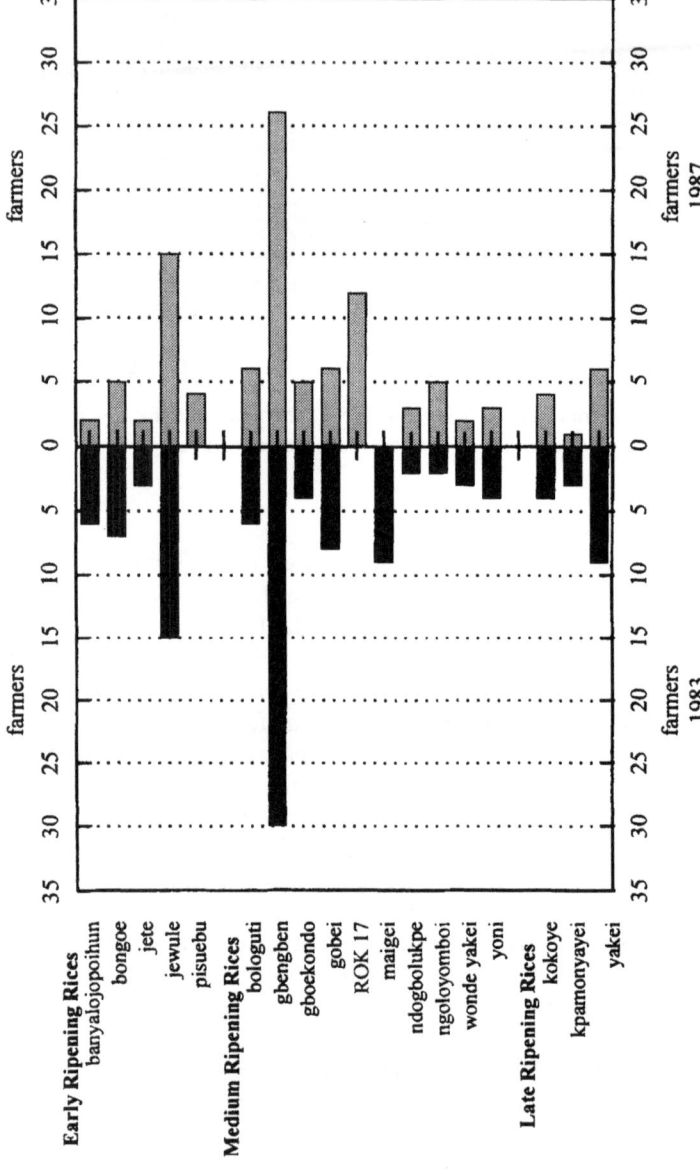

Figure 5. Mogbuama rice varieties in 1983 and 1987: plantings of 52 farmers compared – numbers of farmers.

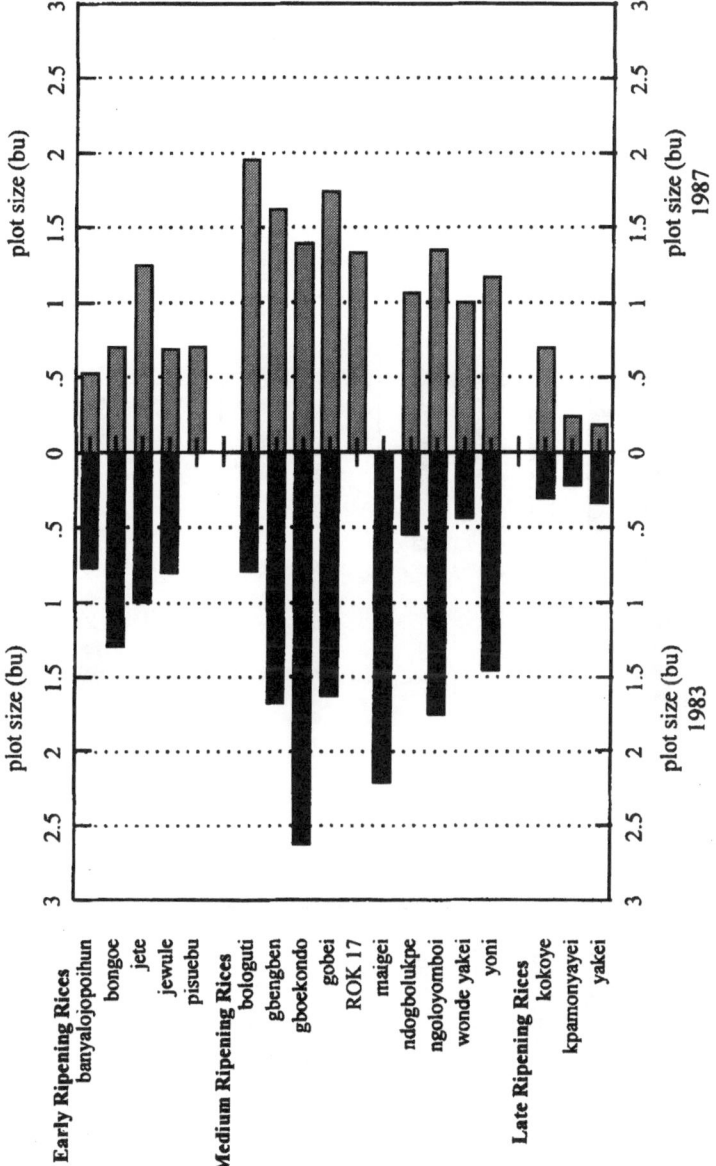

Figure 6. Mogbuama rice varieties in 1983 and 1987: plantings of 52 farmers compared – plot sizes (measured as bushels of rice planted).

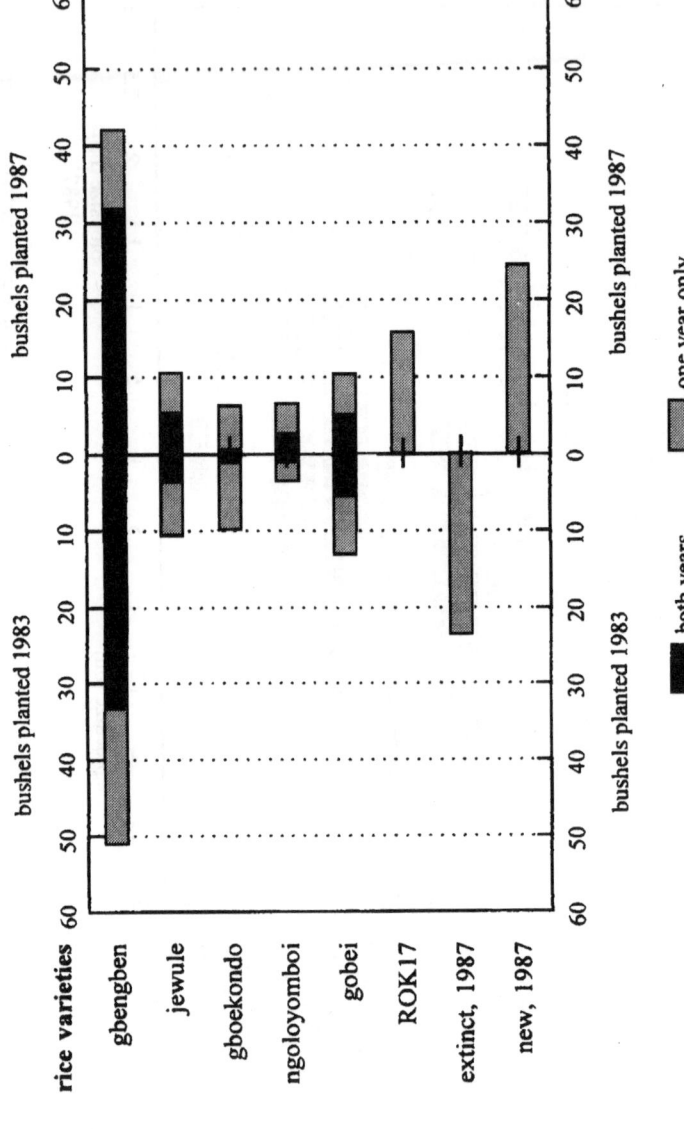

Figure 7. Turnover of selected rice types, Mogbuama, for 52 farmers working in 1983 and 1987.

become too old to farm when lists of active farmers for 1983 and 1987 were compared.

But variety change was also a feature of stable households (i.e., farm units with the same head in 1983 and 1987). The analysis below is based on data for 52 households, out of a total of 59 such stable units. Excluding rices grouped under a generic term **yaka** (i.e., long-duration swamp rices mainly cultivated by household dependents, thus "charity rice," cf. Arabic **zakat**, "alms"), members of stable households planted a total of 37 named varieties in 1983, including 24 four-month upland types and four three-month varieties, and 32 varieties in 1987, including 19 four-month upland types and five three-month varieties (Figures 5 and 6). The upland rice **gbengben** was the most important variety, accounting for 32% and 29% of the total area planted in 1983 and 1987 (Figures 5 and 6). **Jewule** was in both years the most important three-month rice, accounting for roughly half of all three-month rice planted.

The main changes in variety use are summarized in Figure 7. Rices newly adopted since 1983 occupied 17% by area of all rice land in 1987. Conversely, about 15% of land in 1983 was planted to varieties abandoned by 1987. Perhaps even more striking as an index of change, however, is the variety turn-over rate assessed farmer-by-farmer. Farmers using the same variety in 1983 and 1987 accounted for only 46% of the total rices planted in 1987. Expressed as a probability, each household had a greater than 50:50 chance of changing one or more rice varieties in the five-year study period.

Why did so many Mogbuama farmers change rice varieties in this short interval? Three main sets of factors are involved:

First, some changes were involuntary. Rice intended as seed was sometimes eaten due to severity of the hungry season or when a head of household was absent from the village or ill and unable to supervise farm activities. Mogbuama farmers replace lost seed rice by taking loans-in-kind (*lonei hou*) from better-supplied neighbors or kin. But they sometimes have to be content with what they are offered, i.e., an unfamiliar, less-attractive, or badly mixed variety. Such seed may have unanticipated benefits, however. Separated, tested, and multiplied, off-types may suit problematic soils or new and unfamiliar plots, and for this reason, indebtedness is not necessarily a disaster. To the Mogbuama rice farmer, loans-in-kind have some incidental attractions not unlike gambling (Richards 1986). Gambling is, at times, the reason it was necessary to borrow seed in the first place.

Second, some farmers engage in seed exchanges in order to recover a particular variety tried in the past (perhaps lost in a drought) or to replace mixed with unmixed seed. Better a clean bushel of **ndo-gbolukpe**, say, than a mixed bushel of **gbengben**. Badly mixed seed disrupts harvesting calculations since mixed varieties often do not ripen together. Successful seed exchange requires a friend or relative with a surplus of the sought-after variety. Spare and preferred varieties are swapped on a bushel-for-bushel basis. The surplus variety is sold or eaten. Although not all upland rices taste or cook the same, taste generally becomes an issue only when swamp MVs are involved. Some swamp varieties are considered so poor in milling, cooking, or keeping properties that the supplier asks for a differential before agreeing to an exchange, e.g., 1.5 bushels of **CP4** for one of **gbengben**.

Third, farmers give out or beg small amounts of seed, a few handfuls at a time, specifically to run trials (Mende *hungoo*, "to test" or "to look into"). When a farmer experiments with a new or unfamiliar variety, a process often continued for several seasons in succession, it is considered a good idea to send small amounts of the seed to friends and relatives, partly for interest and friendship's sake but also to minimize the risks of failure. Such transactions, especially between affines, from a son-in-law to a mother-in-law, for example, spreads risks across different land-holding groups, villages, and districts. If your own experiment goes wrong your mother-in-law may have better luck, and she will resupply you. Experimentation is also applied to accidental introductions, for example, varieties spread by birds, in elephant droppings, and, more recently, by the sacking in which white rice has been imported, and to off-types and spontaneous crosses rogued during harvest. Whereas loans and exchanges are the business of heads of households, and hence largely male preserves, rice experiments are undertaken by men and women, young and old alike, and women are frequently credited as sources of interesting new varieties.

The rices planted in 1987 in Mogbuama were acquired in the following ways (157 cases). The single most important means was by gift (37%), typically in small amounts for experimentation. Thirty-two percent was acquired by purchase, 18% by exchange, and 10% through loans-in-kind, this last including four instances in which farmers first acquired a variety by taking it in *repayment* of a loan. Most acquisitions had been made since 1983 (70% of cases). In only 13 cases (8%) had the varieties in question first been acquired more than 10 years ago. This figure further supports the picture of high varietal turnover as a normal and enduring feature of Mogbuama rice farming systems. It ought to be noted, however, that not all farmers were reporting first-time acquisitions. In some cases they were describing reacquisitions of lost materials.

The Mogbuama study included the monitoring of an MV (**ROK 17**, formerly **LAC 23**) first introduced as part of the 1983 fieldwork. This pure-line selection from Liberia was well-received and rapidly disseminated because farmers quickly recognized it for what it was, i.e., medium-ripening, four-month upland rice. By 1987, **ROK 17** was second in importance in its class only to **gbengben**, judged by numbers of farmers planting it (Figure 5). Data for average plot size (Figure 6) and total bushels planted (Figure 7) further suggest its growing importance to farmers.

Farmers told me, however, that they were keenest to experiment with three-month varieties. These varieties can be planted in run-off plots at the very beginning of the rains and so contribute to solving the problem of the pre-harvest "hungry season," when rice is in short supply, expensive, and needed to feed farm laborers. For a three-month rice type, earliness, drought tolerance, and some resistance to bird damage are more important characteristics than high yield potential. As of 1987 there were no three-month MV releases suited to local needs. Instead, to secure early varieties suited to their conditions, Mogbuama farmers actively screened off-types, spontaneous outcrosses, and adventitious introductions. Over time, older three-month varieties such as **nduliwa** and **jete** had been discarded in favor of more recent and slightly quicker varieties such as **gete** and **pla-kongoe**. These, in turn, had been displaced by quicker but tougher varieties such as **jewule**. Still the main three-month rice in Mogbuama in 1987, cultivated by 15 households, **jewule** was beginning to be challenged by **pisuebu**, which was not present at all in Mogbuama in 1983 but planted by five households in 1987. Some farmers thought that **pisuebu** would eventually displace **jewule**.

This indigenous screening and selection process for early rices was then applied to a one kilogram packet of the "**Three-Month**" rice acquired in Kamba (Kambia District). By 1992, this small-grained glaberrima type was widely grown in Mogbuama and had become popular in villages up to 25 km distant. That year a bulk purchase of this seed type was made in Mogbuama for the benefit of relief and rehabilitation activities in the war-affected Pujehun District.

Farmer-First Breeding: Germplasm Evidence

As outlined, despite pressure for conformity with Green Revolution breeding ideals, Rokupr Rice Research Station pursued its own low-input selection and breeding strategy from the 1960s. Seemingly, this had an undramatic but not insignificant impact on rice cultivation in Sierra Leone, and more impact than the Green Revolution semi-dwarf strategy, because it fitted well with what a majority of farmers were already trying to do for themselves when making seed choices based on locally available plant resources. Evidence for congruence between breeders' and farmers' concerns is to be found in comparing the IV and MV samples collected from farmers' fields during the 1987–88 surveys.

MV and IV samples were planted and grown under standard conditions on one of two observation sites at Rokupr (depending on whether samples were considered by farmers to be upland or wetland types). The Green Revolution rice ideotype stresses semi-dwarf stature (70–80 cm) and earliness (90–100 days). Farmers in Sierra Leone are keener on taller types and support a mix of duration classes to suit variable soil moisture and topographical conditions. How do farmer-managed MVs compare with IVs in duration and height?

First, we look at earliness. In the upland sample (Figure 8), MVs "track" the IV distribution closely, but with one major exception. This is that MVs contributed nothing at all to the short-duration tail (below 120 days) of what farmers were actually growing in 1987. Perhaps the main "gap" in the Rokupr farmer-first breeding program, early upland rices have begun more recently to attract breeders' attention. To some extent it is a need already met through continued retention of glaberrima varieties (Figure 9). The distribution of wetland MVs likewise "tracks" the pattern of IV plantings well, with a peak at 165 days, which is close to the high point on the IV curve (Figure 10). MVs apparently extend farmers' choices at the extremes of the scale.

Second, we consider height and duration together. Under Green Revolution assumptions we would expect scattergrams of IV and MV height-duration characteristics to be distinct. Under farmer-first assumptions the height-duration scatters will tend to coincide. Figures 11 and 12 confirm that MVs "sit" well within the IV scatter for height and duration in mangrove and upland ecologies. Only in the bolilands (seasonally inundated grasslands) does there appear to be any substantial discrepancy between IVs and MVs. In this case, boliland MV adoptions (dominated by releases such as **CP4**) tend to be found at the

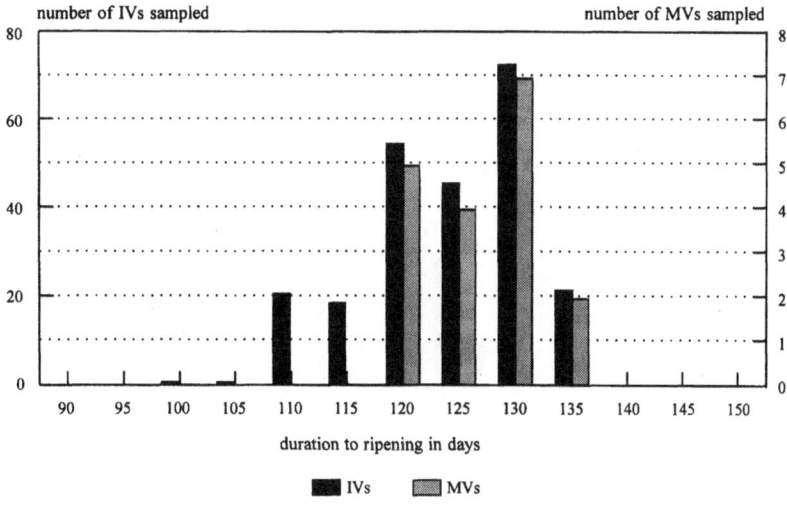

Figure 8. Upland seed samples grown at Rokupr, 1988: indigenous and modern varieties compared – duration to ripening.

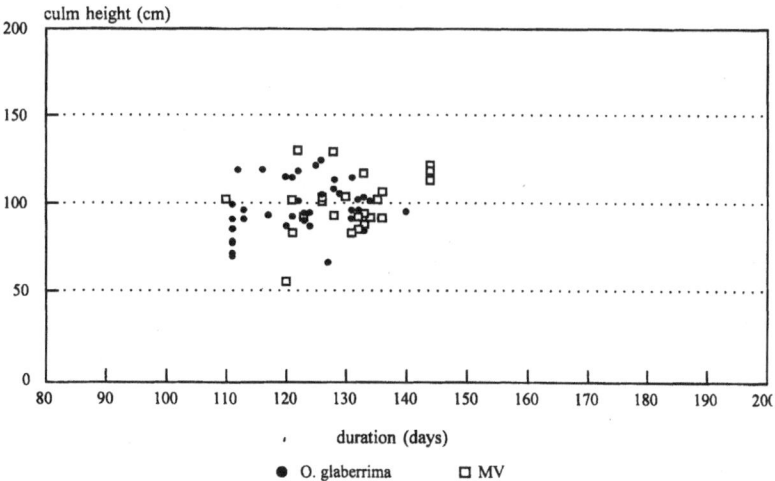

Figure 9. Upland glaberrima and MV types: farmer seed samples, 1987.

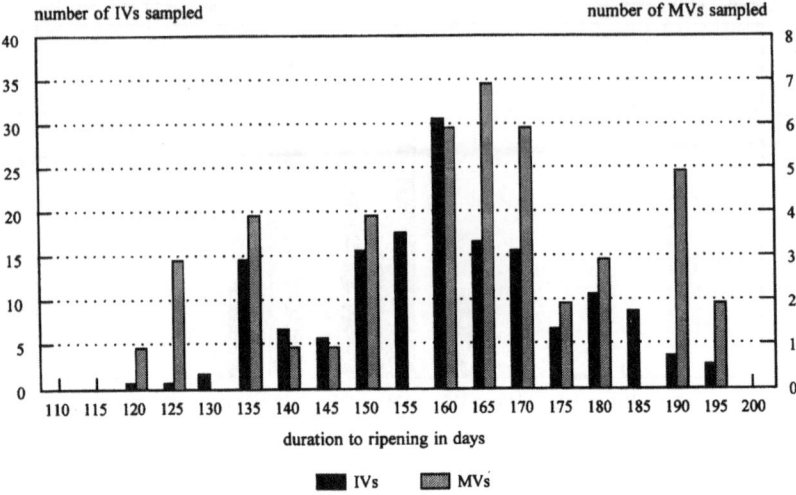

Figure 10. Lowland seed samples grown at Rokupr, 1988: duration to ripening.

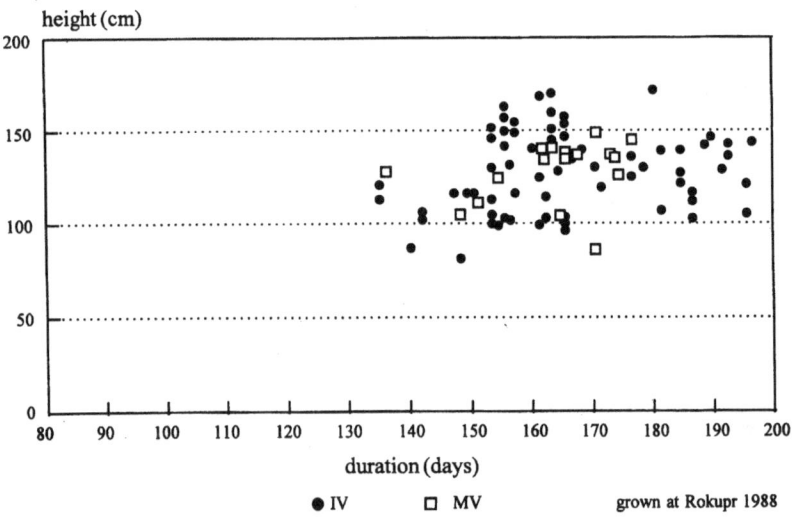

Figure 11. Mangrove IV and MV types: farmer plantings, 1987.

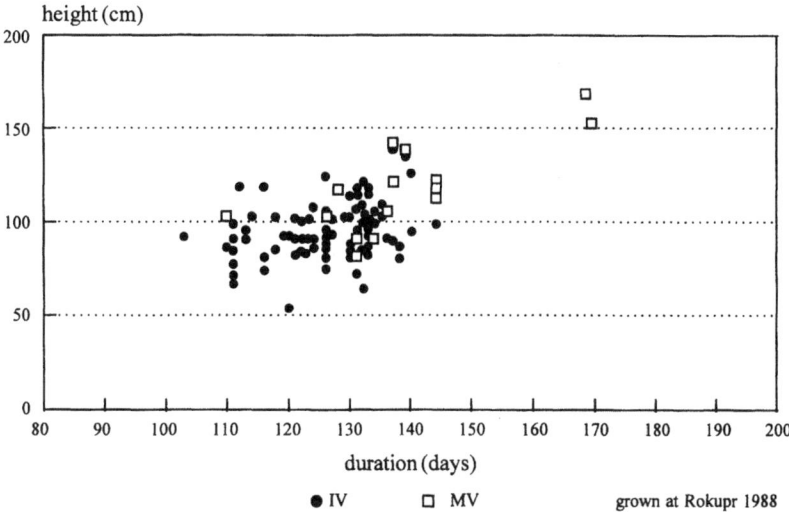

Figure 12. Upland IV and MV types: farmer plantings, 1987.

longer/taller end of the scatter of what farmers actually grow (Figure 13). On balance, the findings provide clear evidence that MV selection and adoption in Sierra Leone has followed a farmer-first rather than Green Revolution path.

The National Seed Board approved a new series of Rokupr rice releases (**ROK 17** to **ROK 32**) for the 1990s. It is interesting to ask where these fit from a farmer-first perspective. Typical height and duration characteristics have been published by the Seed Multiplication Project (SMP). In height-duration terms, three new upland cultivars (**ROK 18–20**) ought to be welcome since they are around 90–100 cm tall (typical for upland IVs) and below 120 days in duration (filling an obvious "gap" in MV provision for uplands). Two mangrove selections (**ROK 21** and **ROK 22**) are suitable in height, but they do not meet farmers' desire for a tall variety that is quicker than **ROK 10** (the niche currently occupied by IVs **fatu yando** and **kolma**). Boliland release **ROK 30** is similar to **CP4** in height and duration, but **ROK 29** (at 140–45 days), being similar in height to, but slightly quicker than, IV **mara**, may prove especially useful. It is less easy to pronounce on inland valley swamp releases. Without water control, farmers tend to favor tall varieties (**ROK 23** and **ROK 26–8** all seem suitable in this respect) and find long duration an advantage in programing a mix of upland and inland valley swamp cultivation (**CP4**,

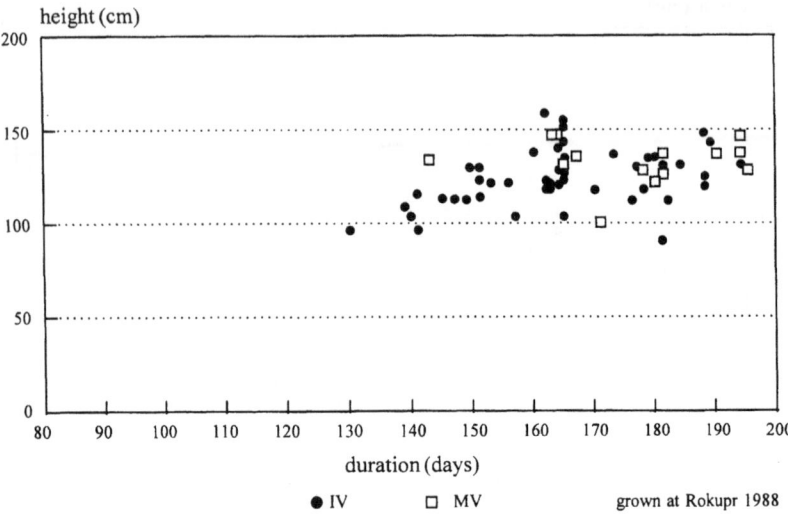

Figure 13. Boliland IV and MV types: farmer plantings, 1987.

for example, is favored as a long-duration wetland variety that can be harvested after all work on uplands is complete). Boliland farmers might favor 150–160 day varieties for their own inland valley swamp conditions, and **ROK 28** might prove a useful short-duration flood-tolerant tall type for farmers moving from upland to inland valley swamps. Only **ROK 20** and **ROK 31** can be considered true Green Revolution semi-dwarf types. At 130 days **ROK 31** might prove a suitable replacement for blast-susceptible **CCA** in lowland farms with good water control.

SOME CONCLUSIONS

The findings of the fieldwork may be summarized as follows:

● The main impact of rice research in Sierra Leone has been farmer uptake of tall, tough, wetland and upland varieties such as **CP4**, **ROK 3**, and **ROK 10**. Surveys produced no *direct* evidence of any significant adoption of semi-dwarf varieties. (Note: MV material for which farmers had no satisfactory name, acquired from projects and called, e.g., "**packet rice**" or "**IDA**," after the source of World Bank funding for IADPs in Sierra Leone, proved in all cases to be non-semi-dwarf material – see Figure 14.)

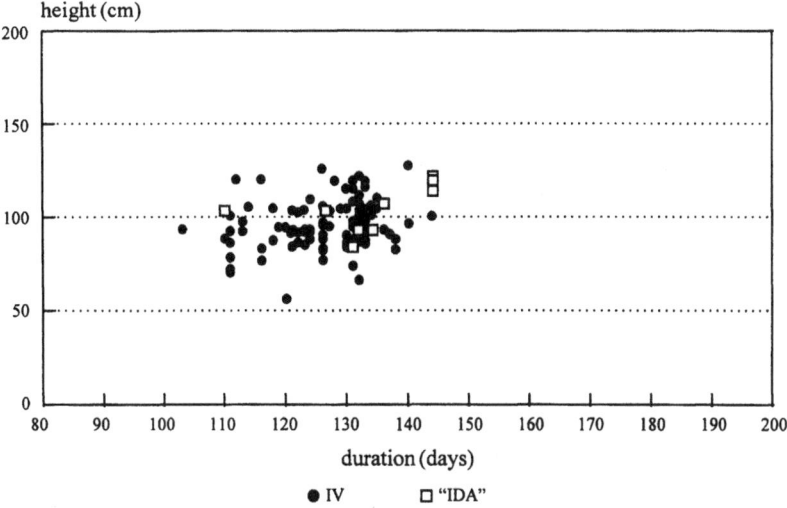

Figure 14. Upland IV and "**IDA**" Types: farmer plantings, 1987.

- Generalizing from the survey data, it would appear that perhaps 15–20% of the national rice hectarage in 1987–88 was accounted for by MVs, with non-Green Revolution MVs being by far the more important group.
- Uptake of MV material in Sierra Leone varies considerably by region and ecology, approaching 40% of all rice planted in the *boli* grassland region, but less than 5% of forested upland-wetland farmland around Njala in central Sierra Leone. (Note: this is despite Njala's role in promoting intensive adaptive research and extension over a number of years.)
- Although projects report optimistic figures for MV adoption in Sierra Leone, there are few reliable independent data against which our own surveys could be cross-checked. In one case where direct comparison was possible, for the adoption of MV types in the mangrove zone, the 1987–88 figure for farmer uptake of MV material – 5% of all mangrove rice planted – matches closely figures from Adesina and Zinnah (1993). (Note: they report sharp rises in MV usage in the years 1989–92.)
- Farmers mainly acquire new rice varieties through informal seed exchange and local credit and purchase arrangements. The local seed system is the main source of supply, even including MV types.

This suggests that MV releases in Sierra Leone are thoroughly absorbed into, and have found their niche within, indigenous farming systems.

- Seed characteristics, and farmer rational response to agro-ecological conditions, seem more important in explaining adoption and abandonment decisions concerning rice in Sierra Leone than the social background of farmers (cf., Adesina and Zinnah 1993).
- The main impact of breeder activity in Sierra Leone has been to enhance the adaptive capacity of farmers to cope with difficult and dynamic conditions. There may be further scope to strengthen this synergistic response by paying even closer attention to farmer repertoires of planting materials and to identifying "gaps" in existing repertoires (e.g., quick-ripening, drought-tolerant upland types needed to deal with rainy-season food shortages).
- The Green Revolution has failed. An alternative is to build further upon the practical, farmer-first, breeding approach of G. S. Banya and others at Rokupr over the last 25 years. Post-Green Revolution research in a poor, war-blighted, country like Sierra Leone may now need to make a decisive break with the concept of "yield *potential*." Emphasis must now be placed on a true anthropology of practice, that is, new rice types should be assessed from the perspective of *actual* performance in *real* (high-stress) farming systems.

POSTSCRIPT: POST-GREEN REVOLUTION RESEARCH, POST-COLD WAR WORLD

Africa was a major site of Cold War struggle. With the collapse of Eastern European Communism in 1989 the world changed. Warfare continues to proliferate in Africa, but it is now no longer structured by Super-Power rivalry. This has caused those who once planned and managed this process of global conflict to speak of the spread of "anarchy" within Africa (cf. Luttwak 1995). Perhaps there is genuine despair that this worldview no longer makes any sense, or perhaps rich countries simply seek to wash their hands of a process, started within the Cold War period, that is now beyond control. In seeking to "naturalize" recent events in Africa proponents of the "anarchy" argument are prone to suggest a vicious circle linking Malthusian crisis and a traditional propensity for violence and barbarism (Kaplan 1994). This diverts attention from two sets of factors vital to understanding modern Africa, i.e., the juxtaposition of "global" life styles

and extreme rural poverty arising from international commercial involvement in the continent, which is most clearly seen in rural mining
enclaves, and the *sociological* significance of rapid population increase
– specifically, the exacerbated *inter-generational tension* traceable to
the rise in the proportion of young people in the population. In several respects, the insurgency in Sierra Leone (1991–1996) can best be
understood as a despairing attempt to draw international attention to
both these sets of issues. This violence is not an inexplicable reversion
to barbarism and anarchy but a product of jarring local juxtapositions of wealth and poverty and the rising despair of young people
suffering a sense of social exclusion.

This essay will conclude with two related questions: "With what
resources will young people in Sierra Leone build a new constructive
vision of the future?" and "What part might new agricultural technologies play in that vision?" First, however, it is necessary to outline
the broader political context within which the older, Green Revolution vision perished.

The Green Revolution in rice research in Sierra Leone was replete
with ironic Cold War associations. The Taiwanese were the earliest
proponents of the Green Revolution in the country. Under a faintly
"Marxist"-leaning All People's Congress (APC) regime after 1968, the
Taiwanese were replaced by rice technicians from Communist China,
at the same time as the country began to receive funding from the
World Bank for IADPs. IADPs aimed to spread the poorly tested
Taiwanese rice agriculture "model" to the generality of the small-
farming community in time to save the "bottom 40%," if one takes the
rhetoric of IADP documentation seriously, from an evil class of village
money lenders – the monopoly capitalists allegedly driving the peasants into the hands of the Communists. I once interviewed one of
these "usurers." "Was he not afraid that the IADP with its cheap
credit program would drive him out of business?" "No," he answered,
"because I live here, with these people, and will continue to live here
with them once the project has collapsed... life is hard, the risks are
real, the IADP will one day take farmers to court for non-repayment
of loans, and who will they turn to help them out of difficulty at that
point...?" Indeed, ten years later after the project had come and gone
he was still sitting outside his modest store, still offering credit to the
same farmers, still struggling to come to terms with harsh economic
realities in an unforgiving environment. But for a few years we were to
witness one of the world's leading capitalist institutions seeking to put
out of business the only true capitalist in the neighborhood in order to

prevent the emergence of a revolutionary consciousness among the peasantry.

The real source of this rich casket of ironies was the APC regime. The sole political party in a one-party state from 1978, with party cadres trained in East Germany, the APC was a regime with its roots in mining trade unionism, but it had no real commitment to socialism and certainly none at all to radical agrarian causes. The party was much more interested in the mineral bonanza and reshaping the commerce associated with diamond mining to its own advantage. This helps explain why, as Abraham and Sesay (1993) note, the IADP policy of "integrated agricultural development" was completely undermined by a systematically over-valued exchange rate that made imported milled rice artificially cheap. Farmer prices for husk rice, the country's staple, declined in real terms by as much as 63% between 1976 and 1987, the main years of investment in the IADP-based Green Revolution strategy. Not unreasonably, farmers decided the Green Revolution was far too risky and expensive an investment, and they continued to develop their own local adaptive strategies. This type of policy "contradiction" is sometimes explained as "urban bias" (the political need to keep urban workers "sweet" – Lipton 1976). Abraham and Sesay (1993) offer a much more direct explanation for the hidden political bias operating against the Green Revolution in Sierra Leone, despite abundant public rhetoric in its favor. Key members of the government made money out of rice importation, and there was little in the way of an organized agrarian interest to oppose them.

What little source of organized *agrarian* opposition in the country there was tended to be concentrated in the rich cocoa and coffee growing region along the Liberian border. APC president Siaka Stevens understood that the eastern border district of Kailahun was one of the most ardent sources of opposition to his rule. Hiding behind IMF advice to close the grossly inefficient railway that had for 70 years linked Kailahun directly to the capital Freetown, Stevens reorganized the flow of diamond wealth from the eastern diamond districts along a new road passing through central Sierra Leone, the region in which the APC garnered much of its political support, and he sought to punish his enemies in the east by building no road to replace the railway from Kenema up to Pendembu, the main town in Kailahun. One estimate (Abraham and Sesay 1993) suggests that 60% of Sierra Leone's export earnings from agricultural commodities originated in Kailahun District. Increasingly, this agricultural produce was smuggled over the border into Liberia, and political disaffection in Kailahun grew unchecked.

If lack of a road had been intended to punish his political opponents, Stevens later recognized it as a mistake. In a farewell interview broadcast on BBC World Service in 1985 the elderly president noted it as the one decision of his period in power he had come to regret (Abrahams and Sesay 1993). But it was too late. A number of political activists, including students, opposed to the one-party state move in 1977, had been driven out of the country into Liberia, some later finding their way to Libya. Circa 1981 they organized an anti-APC political movement in exile, the Revolutionary United Front (RUF). First fighting alongside Charles Taylor's group in the invasion of Nimba County, Liberia, in 1989, the RUF later targeted Kailahun District as a disaffected isolated region in Sierra Leone in which to trigger a rebellion. Lack of decent roads was a key factor protecting the militarily weak RUF forces from counter-attack by the Sierra Leone army.

APC incompetence in prosecuting the war in the east, where it felt threatened by the dubious loyalty of the local population, cost the regime its grip on power when a group of young army officers from the war front came to Freetown to complain about pay and resources and mounted a successful coup in April 1992. The coup stole the anti-APC thunder of the RUF but had little effect on its slow-burning plans to destabilize the country. Perhaps modelled on Shining Path (minus the Maoism) the RUF pitched its appeal mainly to deracinated youth in the diamond fields of eastern and central Sierra Leone. Diamond tributers were already inured to great hardship and high levels of violence. They come from all over Sierra Leone. They have tasted modernity through part-completed education and modern electronic media, and daily live the jarring juxtapositions of international wealth and poverty, where diamonds may be "forever" but the lives of diamond diggers are of little account. Where the RUF was unable to recruit members it seized them by force. Many captives quickly accepted the conditions of their new existence, since the political vision under the APC, such as it was, had long since perished. Internally, the RUF stresses four visionary themes of its own: a strong stand against corruption, youth empowerment, mass education, and health for all. Like Shining Path it appears to have a highly decentralized structure, offering command to youths and women as well as men, with units apparently free to exercise their own initiative in picking targets.

Driven out of Kailahun at the end of 1993, RUF units regrouped in forest reserves in the center of the country. In some areas, notably Pujehun District, the RUF seemingly operated in conjunction with various local dissident groups. An appearance of anarchy is further heightened by apparent splits within the Sierra Leone military forces, with some disgruntled soldiers, and perhaps whole units, in the war zone taking part in banditry or fraternizing with the enemy for economic gain. Both military personnel and insurgents mined diamonds in lulls in the fighting, and war became endemic on the back of lucrative deals to be struck over diamonds, arms, and drugs (Keen 1995). Political motives accounted for some fraternization and banditry. Until 1992, advancement in the army depended on APC patronage, and acts of military incompetence may, in some cases, have been calculated moves by APC loyalists to undermine the military government of Captain Strasser. Tensions within the army were further exacerbated by extra-judicial executions in December 1992 of popular army officers and leading figures in the APC regime.

Early in 1995 the RUF came within range of the capital Freetown and struck at most parts of the country in a series of lightning raids. Unable to hold any sizeable town against the Sierra Leone army, the RUF focuses mainly on acts of devastation and terror in smaller rural settlements. Offered a place in the political process, the movement refused to respond, partly out of fear for its own safety and partly because, ideologically, it sees itself as being "above" a political process it deems inherently corrupt. The RUF has convinced itself that all international organizations, the World Bank and United Nations as much as international mining interests, are deeply enmeshed in the corruption afflicting Sierra Leonean society. Like Shining Path, it seeks to regenerate society from within, starting where it stands in the forest (Richards 1996).

There seems little prospect that the RUF would make much headway as a popular choice under democratic politics. It has totally destabilized the countryside, and most rural dwellers want little more than to be left in peace to get on with their lives. On the other hand, RUF emphasis on rebuilding from within may not be a totally inapposite standpoint for war-peace transitions in Sierra Leone. The critique of external influence as inherently "corrupting," while aimed primarily at the jarring contradictions of wealth and poverty experienced on a daily basis by young participants in the diamond economy, is also to some extent applicable – as we have seen – to the Green Revolution, at least in its ironic Sierra Leonean manifestation. The idea of a dedicated national class of educated professionals, in this case rice researchers, working directly with a mass clientele on a shared project constructively to manage the local heritage of plant genetic resources has a number of visionary possibilities that might suit it to the inspirational needs of post-war reconstruction. This is but the grain of an idea, with some immediate implications for self-managed rehabilitation of farming populations displaced by war (Richards, Ruivenkamp, van der Drift, Gonowolo, Jusu, and Longley 1995). Some of the science to support a Green Revolution "from below" is already in place (and indeed has been apparent since 1971!), but much hard thinking and ingenuity would now be needed to extend it to the insurgent combatants themselves, since, as conscript diamond tributers and the like, they hardly constitute an agrarian class fraction in any recognizable sense. On the other hand, modern biotechnologies and agricultural hardware technologies are plastic resources, and until sustained attention is paid to the issue of how these young people see their own futures, no one can say for certain that beating swords into plowshares is impossible in the African bush.

END NOTES

1. In this chapter the word "type" is used as a broad term to refer to any rice plant that is recognized by farmers or breeders as both distinctive and recurrent. The term is used where "variety" would be too specific, carrying implications of an officially released pure-line selection. The word "ideotype" refers to the breeder's concept of "plant ideotype" as elaborated by Donald (1968). Ideotype is a design concept – specifically, it is an ideal plant type envisaged by breeders as a target for breeding programs. As such, the ideotype exists in the minds of breeders in advance of its realization in nature. The Green Revolution ideotype was a quick-maturing, short-straw, weakly tillering rice plant, intended to allow plant populations to make better overall use of available space, nutrients, sunlight, and moisture on a restricted area of land. In reading this essay, the different meanings of the terms "type" and "ideotype" should be kept clearly in mind.

2. Seedlessness, rather than landlessness, is probably the defining characteristic of the very poor in rural Sierra Leone. Indeed, acquiring enough rice to plant a farm each year is a major concern for most small-scale farmers in the country. It is not uncommon to encounter areas of cleared land abandoned due to lack of seed. The problem is greatest for the very poor in years following poor harvests caused by drought or early rainfall, and, since 1991, by the dislocations of war.

REFERENCES

Abraham, A., and H. Sesay
 1993 Regional Politics and Social Services Provision Since Independence. *In* The State and the Provision of Social Services in Sierra Leone Since Independence, 1961–1991. C. Magbaily Fyle, ed. Dakar: CODESRIA.

Adesina, A., and M. Zinnah
 1993 Impact of Modern Mangrove Swamp Rice Varieties in Sierra Leone and Guinea. International Rice Research Notes 18(4):36.

Borlaug, N. E.
 1972 Breeding Wheat for High Yield, Wide Adaptation, and Disease Resistance. *In* Rice Breeding. Pp. 581–589. Los Baños, Philippines: International Rice Research Institute.

Boyer, P., ed.
 1992 Cognitive Aspects of Religious Symbolism. Cambridge: Cambridge University.

Chambers, R., A. Pacey, and L. A. Thrupp, eds.
 1989 Farmer First: Farmer Innovation and Agricultural Research. London: Intermediate Technology.

Chandler, R. F.
 1972 The Impact of the Improved Tropical Plant Type on Rice Yields in South and Southeast Asia. *In* Rice Breeding. Pp. 77–84. Los Baños, Philippines: International Rice Research Institute.
 1982 An Adventure in Applied Science: A History of the International Rice Research Institute. Los Baños, Philippines: International Rice Research Institute.

Donald, C. M.
 1968 The Breeding of Crop Ideotypes. Euphytica 17:385–403.

Douglas, M., and A. Wildavsky
 1982 Risk and Culture: An Essay on the Selection of Technological and Environmental Dangers. Berkeley: University of California.

Fujimura, J.
 1992 Crafting Science: Standardized Packages, Boundary Objects, and "Translation." *In* Science as Practice and Culture. A. Pickering, ed. Chicago: University of Chicago.

Glanville, R. R.
 1938 Rice Production on Swamps. Sierra Leone Agricultural Notes, No. 7.

Huang, C. H., W. L. Chang, and T. T. Chang
 1972 *Ponlai* Varieties and Taichung Native 1. *In* Rice Breeding. Pp. 31–45. Los Baños, Philippines: International Rice Research Institute.

IRRI (International Rice Research Institute)
1972 Rice Breeding. Los Baños, Philippines: IRRI.

Jusu, M. S.
1995 The Genus *Oryza*: Sources and Uses of Genetic Variability by Susu and Limba Farmers in Sierra Leone. Unpublished research proposal, Working Group for Technology and Agrarian Development and Department of Plant Breeding, Wageningen Agricultural University.

Kaplan, R. D.
1994 The Coming Anarchy: How Scarcity, Crime, Overpopulation, and Disease Are Rapidly Destroying the Social Fabric of Our Planet. Atlantic Monthly. February, 1994, pp. 44–76.

Karimu, J. A.
1981 Strategies for Peasant-Farmer Development: An Evaluation of a Rural Development Project in Sierra Leone. Ph.D. thesis, University of London.

Karimu, J. A., and P. Richards
1981 The Northern Area Integrated Agricultural Development Project: The Social and Economic Impact of Planning for Rural Change in Sierra Leone. Research Papers, Department of Geography, School of Oriental and African Studies, University of London.

Keen, D.
1995 "Sell-Game": The Economics of Conflict in Sierra Leone. Paper presented at a one-day conference, "West Africa At War: Anarchy or Peace in Liberia and Sierra Leone?" held at the Department of Anthropology, University College London, UK, October 21, 1995.

Lawson, E. T., and R. N. McCauley
1990 Rethinking Religion: Connecting Cognition and Culture. Cambridge: Cambridge University.

Latour, B.
1993 We Have Never Been Modern. Hemel Hempstead: Harvester Wheatsheaf.

Lipton, M.
1976 Why Poor People Stay Poor: Urban Bias in World Development. London: Temple Smith.

Luttwak, E. N.
1995 Great Powerless Days. Times Literary Supplement. June 16, 1995, pg. 9.

McDonald, D. J.
1972 Comment in discussion on "Varietal Responses to Some Factors Affecting Production of Upland Rice." *In* Rice Breeding. Pg. 700. Los Baños, Philippines: International Rice Research Institute.

Mackenzie, D.
    1992   Economic and Sociological Explanations of Technical Change. *In*
            Technological Change and Company Strategies. R. Coombs, P. Sav-
            iotti, and V. Walsh, eds. London: Academic.

Moore-Sieray, D.
    1988   The Evolution of Colonial Agricultural Policy in Sierra Leone with
            Special Reference to Swamp Rice Cultivation, 1908–1939. Ph.D. the-
            sis, University of London.

Richards, P.
    1985   Indigenous Agricultural Revolution: Ecology and Food Production
            in West Africa. London: Hutchinson.
    1986   Coping with Hunger: Hazard and Experiment in a West African
            Farming System. London: Allen and Unwin.
    1995   The Versatility of the Poor: Wetland Rice Farming Systems in Sierra
            Leone. Geoforum 35(2):197–203.
    1996   Fighting for the Rain Forest: War, Youth and Resources in Sierra
            Leone. London: James Currey for the International African Institute.

Richards, P., and G. Ruivenkamp
    1996   New Tools for Conviviality: Society and Biotechnology. Unpub-
            lished ms.

Richards, P., G. Ruivenkamp, R. van der Drift, M. Gonowolo, M. Jusu, and
C. Longley
    1996   Seeds and Survival: Crop Genetic Resources in War and Recon-
            struction in Africa. Report for the International Plant Genetic Re-
            sources Institute, Rome. Forthcoming.

Roberts, L. M.
    1972   Prospects for the Future. *In* Rice Breeding. Pp. 715–721. Los Baños,
            Philippines: International Rice Research Institute.

Index